Biosystems & Biorobotics

Volume 30

The BIOSYSTEMS & BIOROBOTICS (BioSysRob) series publishes the latest research developments in three main areas: 1) understanding biological systems from a bioengineering point of view, i.e. the study of biosystems by exploiting engineering methods and tools to unveil their functioning principles and unrivalled performance; 2) design and development of biologically inspired machines and systems to be used for different purposes and in a variety of application contexts. In particular, the series welcomes contributions on novel design approaches, methods and tools as well as case studies on specific bio-inspired systems; 3) design and developments of nano-, micro-, macro- devices and systems for biomedical applications, i.e. technologies that can improve modern healthcare and welfare by enabling novel solutions for prevention, diagnosis, surgery, prosthetics, rehabilitation and independent living. On one side, the series focuses on recent methods and technologies which allow multi-scale, multi-physics, high-resolution analysis and modeling of biological systems. A special emphasis on this side is given to the use of mechatronic and robotic systems as a tool for basic research in biology. On the other side, the series authoritatively reports on current theoretical and experimental challenges and developments related to the "biomechatronic" design of novel biorobotic machines. A special emphasis on this side is given to human-machine interaction and interfacing, and also to the ethical and social implications of this emerging research area, as key challenges for the acceptability and sustainability of biorobotics technology. The main target of the series are engineers interested in biology and medicine, and specifically bioengineers and bioroboticists. Volume published in the series comprise monographs, edited volumes, lecture notes, as well as selected conference proceedings and PhD theses. The series also publishes books purposely devoted to support education in bioengineering, biomedical engineering, biomechatronics and biorobotics at graduate and post-graduate levels.

Indexed by SCOPUS, WTI Frankfurt eG, SCImago

More information about this series at https://link.springer.com/bookseries/10421

María Amparo Grau Ruiz
Editor

Interactive Robotics: Legal, Ethical, Social and Economic Aspects

Selected Contributions to the INBOTS Conference 2021, 18–20 May, 2021

Springer

Editor
María Amparo Grau Ruiz
Visiting Professor of Transnational Taxation
Northwestern University
Evanston, USA

Full Professor of Financial and Tax Law
Complutense University of Madrid
Madrid, Spain

ISSN 2195-3562 ISSN 2195-3570 (electronic)
Biosystems & Biorobotics
ISBN 978-3-031-04304-8 ISBN 978-3-031-04305-5 (eBook)
https://doi.org/10.1007/978-3-031-04305-5

This Springer imprint is published by the registered company Springer Nature Switzerland AG
The registered company address is: Gewerbestrasse 11, 6330 Cham, Switzerland

Dedicated to Our Fathers

Foreword

The history of robotics is one of ever closer interaction with humans. While robots were originally only intended for use in industrial environments to replace humans in dangerous, tedious and repetitive tasks and tasks requiring precision, the current scenario is one of transition toward increasing interaction with the human operator, and where robots are a form of embodied artificial intelligence (embodied AI). This means that interaction with humans is expanding from a mere exchange of information (for instance, in teleoperation tasks and service robotics) to a close interaction involving physical and cognitive components, such as the one found in collaborative robots or cobots.

As a result of international cooperation, the adoption of artificial intelligence (AI), data and robotics—and their integration in embodied AI—as well as the process of deepening their applications to provide increased functionality and address societal challenges are being promoted. Critical to accelerate innovation and the "third expansion wave" of embodied AI in more application scenarios is the identification of tractable challenges, the reduction of hurdles in the innovation path, and the promotion of acceptance and understanding by everyone. As the result of analyzing the ethical, legal and socio-economic context for embodied AI under the leadership of Prof. Grau Ruiz, this book—resulting from the European Commission Coordination and Support Action "Inclusive Robotics for a better Society, INBOTS"—is definitely a giant leap forward in shedding light onto the societal framework where this revolution is happening.

The unique values provided by robotics come from the composition of technologies and the embedded integration of AI. The need to develop physically embodied AI systems to interact with persons in unknown and unstructured environments makes robotics intrinsically complex and different, but potentially a ubiquitous and pervasive technology. However, opportunities rarely come without potential risks, which need to be elucidated and contained. The rapid advances in robotics will make it difficult for the public to completely understand the state of the technology, which creates negative perceptions and overestimated expectations, a matter that Prof. Grau Ruiz and colleagues contribute clarity to. Embodied AI should aspire to be a robotics of the people, by the people and for the people. This involves the consideration of a range of issues regarding autonomy, dependence, vulnerability, functioning, care and disability, again from ethical, political and policy perspectives.

What is more, robotics will not only affect individuals and the innovation framework, it will also directly affect society at large from many different perspectives. There are growing interest and international consensus about the urgent need of discussing the implications of deploying highly capable robotic technologies and how this technology is disrupting our current legal and economic paradigms. In this context, questions such as what the main legal, ethical and socio-economic aspects for a proper inclusion of robots into society are, how to develop a robotic responsible research and innovation paradigm, how to assess the sound use of robots in line with sustainable development goals or how to monitor and integrate ongoing, worldwide efforts in legislation need to be carefully addressed. Successfully doing so will help identify the societal values, needs and expectations of embodied AI, define a common strategy to increase public awareness on new robotics avenues, which will promote acceptance of robotics in society and the connection between robotics development and societal development.

This book brings together a comprehensive discussion by leading experts on Ethics, on Labor, on Economy, Intellectual Property and Taxation, on Corporate Social Responsibility and Sustainability, and on Governance & Regulation relating to embodied AI and robotics. The book is particularly comprehensive in providing a worldwide overview of financial tax law, labor and social security law. Inclusive embodied AI can only be achieved if the robotics industry and its researchers establish unconventional collaborations with researchers from other disciplines, such as philosophers, lawyers, sociologists, anthropologists, medical experts and economists. This book is a perfect example of this unconventional, yet cross-fertilizing, collaboration among these and other disciplines. I firmly believe that the process of bringing inclusivity to robotics and embodied AI will position the field as an important contributor to the construction of a fairer society.

Jose L. Pons

Preface

The session "Debate on legal, ethics & socio-economic aspects of interactive robotics" was held during the Final INBOTS Conference in May 2021, organized by the Consejo Superior de Investigaciones Científicas–CSIC (the full program can be found at http://inbotsconference2021.inbots.eu/sessions/). In the framework of the European Union Horizon 2020 project "Inclusive Robotics for a Better Society", Work Package 2, led by Universidad Complutense de Madrid–UCM, arranged a workshop designed to disseminate the results of this project on ELSE topics with the participation of partners in the consortium (http://inbots.eu), a number of whom were part of the recent EU high-level expert groups and connected to EU Robotics. Additionally, prestigious speakers from relevant organizations were invited, such as the International Labor Organization, the International Organization of Employers and the DG Taxud (the full list is available at http://inbotsconference2021.inbots.eu/plenary-speakers/). After making a specific call for papers, an impressive number of contributions were received. The majority of the contributions pointed out critical aspects in the field of interactive robotics from the perspective of social and human sciences. All these contributions were peer-reviewed by ethics and law experts, and a set of selected papers is now published as a book of proceedings. They clearly reflect the current state of the art and demonstrate the critical need for further reflection on how human beings are dealing with the design and deployment of robots and other related technologies.

This volume continues the series started with Inclusive Robotics for a Better Society, Selected Papers from INBOTS Conference 2018, October 16–18, 2018, Pisa, Italy. This book gathers views on robotics from very different academic perspectives and various national backgrounds. It is born after a truly interdisciplinary and international dialogue. It illustrates that the first hesitant steps to greater legal certainty require solid ethical foundations. Moreover, it also highlights the need to take into account the Sustainable Development Agenda and the crucial role of a responsible business conduct adopted by robotics companies. I strongly believe that taking stock of all these ideas will serve for a sound evolution in the future.

I want to express my gratitude to Prof. Dr. Jose L. Pons for having invited me and my UCM colleagues to participate in INBOTS, and for writing the foreword as well. His clear vision on robotics led him to conceive this CSA project and coordinate it in the first period to achieve its mission. Also, to his team at CSIC, for having supported our research. Now that is has finished, we are and will be continuing our work through new projects (like AudIT-S PID2019-105959RB-I00 https://www.ucm.es/proyecto-audit-s/).

I sincerely appreciate the invaluable help of Dr. Fiachra O'Brolcháin, Alina Alexandra Risser, Virginia Martínez Torres and Álvaro Falcón Pulido in the reviewing and editing process. Also of Dr. Leontina Di Cecco and Jayarani Premkumar.

Finally, thank God, I want to recognize my family and good friends. They are always there helping me with my aspirations to make as much as we can to build a better world for everyone, which is more necessary than ever in these strange times.

María Amparo Grau Ruiz

Organization

Committees

Organizing Committee of the INBOTS Conference 2021

Juan C. Moreno	CSIC, Spain
Pilar Raya	CSIC, Spain
Roberto Conti	IUVO, Italy
María Amparo Grau Ruiz	Complutense University of Madrid, Spain
Monica Malvezzi	University of Siena, Italy
Saskia Maresch	DIN, Germany
Andrea Bertolini	SSSA, Italy
Britt Östlund	KTH, Sweden
María Prieto	Inncome, Spain
Garbiñe González	Inncome, Spain

Organizing Committee of the WP2 Session on ELSE Issues

María Amparo Grau Ruiz	Complutense University of Madrid, Spain
Yolanda Sánchez-Urán Azaña	Complutense University of Madrid, Spain
José Ignacio López-Sánchez	Complutense University of Madrid, Spain
Txetxu Ausín	CSIC-Institute of Philosophy, GEA, Spain
Maarten Goos	University of Utrecht, The Netherlands
Rita de la Feria	University of Leeds, United Kingdom
Vincent Müller	University of Leeds, United Kingdom
Fiachra O'Brolchain	Dublin City University, Ireland
Luke McDonagh	City University of London-London School of Economics, United Kingdom

Spanish National Research Council (Consejo Superior de Investigaciones Científicas, CSIC). The Neural Rehabilitation Group at Cajal Institute coordinates INBOTS CSA

Iuvo S.r.l.

Complutense University of Madrid (UCM). IUS-SustentaRSE, GIPTIC and Human Resources Groups

University of Siena, Human Centered Robotics Group

German Institute for Standardization (DIN)

SSSA Pisa, DIRPOLIS Institute

Royal Institute of Technology (KTH)

Inncome PKF Attest

Contents

Ethics

Basic Issues in AI Policy

Vincent C. Müller[1,2](✉)

[1] Ethics of Technology, Philosophy and Ethics Group, Faculty of IE/IS, Technical University of Eindhoven (TU/E), Atlas 9.329, De Zaale, P.O. Box 513, 5600 Eindhoven, MB, The Netherlands
[2] IDEA Centre, Faculty of Philosophy, Religion and History of Science (PRHS), University of Leeds, 17, Blenheim Terrace, Leeds LS2 9JT, UK
v.c.muller@leeds.ac.uk
http://www.sophia.de/

Abstract. This short paper summarises some of the basic points of AI ethics and policy as they present themselves now. We explain the notion of AI, the main ethical issues in AI and the main policy aims and means.

1 AI

The term "Artificial Intelligence" (AI) is now used in two main meanings:

(a) AI is a research programme to create computer-based agents that can show complex behaviour, capable of reaching goals [1], OR
(b) AI is a set of methods employed in the AI research programme for perception, modelling, planning, and action: machine learning (supervised, reinforced, unsupervised), search, logic programming, probabilistic reasoning, expert systems, optimisation, control engineering in robotics, neuromorphic engineering, etc. Many of these methods are also employed outside the AI research programme [2–5]

The original research programme (a) from the "Dartmouth Conference" in 1956 onwards was closely connected to the idea that computational models can be developed for cognitive science of natural intelligence and then implemented on different hardware, i.e. on computing machines. This programme ran into various problems in the "AI Winter" ca. 1975–1995 and the word "AI" got a bad reputation; it thus branched into several technical programmes that used their own names (pattern recognition, data mining, decision support system, data analytics, cognitive systems). After 2000, AI saw a resurgence, with faster hardware, more data, and a stress on neural network machine-learning systems. From ca. 2010 "AI" became a buzzword that resonated in circles outside computer science; now everyone wants to be associated with AI. As a result, the meaning of "AI" is currently broadening towards (b). (More details on this development and the contrast to robotics in [6].

2 AI Ethics

As a result of the popularity of AI, around 2015, the possible negative effects of AI on humanity came into focus as well, e.g. whether AI would take human jobs, decide over

M. A. Grau Ruiz (Ed.): INBOTS 2021, BIOSYSROB 30, pp. 3–9, 2022.
https://doi.org/10.1007/978-3-031-04305-5_1

human lives, spell the end of privacy, kill humans with autonomous weapons or just take over the world entirely in a 'singularity'. Eurobarometer surveys showed that European citizens' views on robots had become more negative [7, 8].

Work on *ethics of AI* had existed since the 1950ies but it was a very small minority of researchers who worked on it, mostly in the context of *computer ethics and data ethics* (esp. on privacy) [for a history, see 9]. AI ethics is closely connected to the philosophy of AI. As a field, it is reaching maturity in the present years, and surveys of the state of the art have only appeared since 2020 [6, 10–12].

AI ethics is an application of the methods of ethics, i.e. it is a branch of "applied ethics". Ethics is the classical discipline of philosophy that describes how to evaluate actions as right or wrong, i.e., how to use normative statements that involve values. The basis for such ethical judgment is typically an evaluation of personal virtues (positive character traits, e.g. honesty or friendliness), of ethical rules or duties (e.g. "do not steal") and of the consequences of an action (typically in terms of utility). Ethical *dilemmas* occur when rules conflict (e.g. honesty vs. politeness) or when rules conflict with outcomes (e.g. honesty results in harm for an innocent person). It is clear that a comprehensive theory of ethics, and thus a comprehensive applied ethics, must take into account all three components: the person performing the action, the type of action and the consequences of the action.

Sometimes it has been assumed that if machines act in ethically relevant ways, then we need a machine ethics, for "ethical machines", for machines as subjects, rather than for the human use of machines as objects [13–16]. However, this is very likely a misunderstanding since there is general agreement that current and foreseeable AI systems do not have what it takes to be responsible for their actions (moral agents), or to be systems that humans should have responsibility towards (moral patients) [17]. So, the responsibility remains firmly with the humans and for the humans – as well as other animals.

What is at stake in AI ethics is the human design and use of AI. There seems to be reasonable agreement that the main ethical issues are:

- Privacy & Surveillance
- Manipulation of Behaviour
- Opacity of AI Systems
- Bias in Decision Systems
- Human-Robot Interaction
- Automation and Employment
- Autonomous Systems and Responsibility
- Machine Ethics
- Artificial Moral Agents
- Singularity

For details on each of these, see the 10 sections on the "main debates" in [6].

If successful, AI ethics says what is right and what is wrong, i.e. it defines what "ethical AI" would be like (see below). Knowing what ethical AI is like is a significant input to *AI policy* (see below) because it defines what the main policy *aims* of AI policy should be.

3 Ethical AI

3.1 Explanation

If AI is designed and used ethically, as defined by AI ethics, then it is *ethical AI*. This desirable state has often been expressed in different terms that sound less academic, e.g. using the term "Human-centric AI" is an approach to AI that judges AI by its positive effects on humans, which captures most of what we want from ethical AI. It stresses that AI should serve human well-being, guarantee human rights or dignity, and have positive effects on society. All of these things are traditionally captured in ethics, which takes into account rights and positive consequences (in deontological and consequentialist considerations).

Some years ago, there was talk of developing "trust in AI", especially after the Eurobarometer had shown that there was little trust – focusing on threats to jobs from AI ? and by autonomous robots. However, one can trust a system for good and for bad reasons, so what matters is whether it *deserves* to be trusted, whether it is "trustworthy". Thus "trustworthy AI" became the central term in the EC White paper [18] and the slogan for EC AI policy.

Much speaks in favour of just calling this "good AI", but "ethical AI", "human-centred AI" or "trustworthy AI" will do just as well.

3.2 AI Ethics Guidelines

In the effort to explain what "ethical AI" might mean, in ca. 2015–2020, many institutions published "ethics guidelines" (under several terms) and to explain how they are in favour of ethical AI. There is a map on https://aiethicslab.com/big-picture/, a list on https://inventory.algorithmwatch.org (and http://www.ptai.org/TG-ELS/policy) and there are recent surveys of these [19, 20]. What these guidelines do is usually not very exciting; they look at the points where ethical issue do occur, like the list of 10 above, and say that we should be ethical – rarely going into the points where there is actually a difficulty in saying *what is ethical*, i.e. a dilemma. For example, the Independent High-Level Expert Group on Artificial Intelligence set up by the European Commission [21] says:

- AI should guarantee:
- Human agency and oversight
- Technical robustness and safety
- Privacy and data governance
- Transparency
- Diversity, non-discrimination and fairness
- Societal and environmental wellbeing
- Accountability

This really just says that AI must guarantee responsibility and liability (1 & 7), positive utility in the long run (2 & 6) and safeguard rights like privacy, autonomy and justice (3 & 4 & 5). None of this comes as news to AI ethics.

The question is *what should be done* to make sure ethical AI is a reality. This is where AI policy comes in.

4 AI Policy

4.1 Definition

A policy is a set of principles to guide organisational actions (typically of states or supranational organisations). A policy consists of two main components: policy *aims* and policy *means* to further those aims.

4.2 Policy Aims

In our case of AI, policy *aims* are constituted by a) ethical principles on AI, plus b) general policy aims, such as those formulated in the 17 UN Sustainable Development Goals (https://sdgs.un.org).

It is thus important for AI policy to specify the specific (a) and general (b) aims it pursues.

While there is significant agreement on the specific ethical principles on AI, general policy aims will often differ between nation states, e.g. the US has a high sensitivity to bias and discrimination, China and the US place great emphasis on geostrategic aims (will thus be reluctant to limit automated weapons), the EU will be sensitive to monopolies and places great emphasis on privacy, many states will put a higher emphasis on economic development than on justice, etc.

Which policy aims will *actually be pursued* also depends on powers that the state actors are subjected to, e.g. public opinion, lobbying, technical feasibility, cost, etc.

Setting policy aims, e.g. through AI ethics guidelines, should not be identified with policy, since policy also involves practical activity to achieve these aims.

4.3 Policy Means

Policy *means* are the practical instruments and methods with which we can further the policy aims. It appears that the main bottlenecks of AI policy, at the moment, is with the appropriate policy means.

The first thought of policy aims goes for legal regulation, but it is important to remember the broad range of other options that exist:

- educational efforts (e.g. curriculum of AI degrees)[1]
- framework for legal liability (e.g. insurance)
- impact assessment tools
- legal regulation
- PR measures
- public spending
- self-assessment frameworks
- self-regulation (in industry)
- self-regulation (e.g. a "Hippocratic oath")
- supporting ethics by design
- taxation
- technical standards (in a framework of legal regulation)

There is material that can help in this area and in politics and political science there is ample experience in developing policy & in facing the bottlenecks of practical policy means.

From other cases of ethics-driven policy, there are models that we can follow, e.g. medical ethics [22] or engineering ethics [23], or professional ethics in various areas. What can be learned from traditional computer ethics, esp. data ethics [24, 25]. What are the ways forward in a professional ethics for AI engineers [26]? Are the principles we have "actionable" [cf. 27]? Perhaps self-assessment is the way, such as the ALTAI self-assessment tool (EC, Unit A1; July 2020): https://altai.insight-centre.org [28]. Further to "ethical AI", there is also a movement to use AI to further other social aims, i.e. to use "AI for social good" [29] – this goes beyond AI policy.

4.4 EU Regulatory Efforts

In the EU, a dominant notion for policy means is "risk". The AI HLEG says there are two kinds of risks [21]: Risks for fundamental rights, including personal data and privacy protection and non-discrimination. Risks for safety and the effective functioning of the liability regime. These risks are then classified in terms of high-risk vs. low-risk. (This is a bit misleading since the rights are not at risk, the risk is that the rights could be violated. Typically, one distinguishes *rights* (that exist independently of positive or negative consequences) from *risks* (the probability of a negative consequence). Any ethics will involve both a notion of rights/obligations and a consideration of outcomes.)

This is combined with moves towards certification, ex ante and post hoc, voluntary and involuntary, self-certification and certification through official bodies. (In Germany through the BSI, DIN and VDA [cf. 30]).

Acknowledgments. This work was sponsored by the European Commission under the INBOTS and IA-AI projects.

References

1. McCarthy, J., Minsky, M., Rochester, N., Shannon, C.E.: A proposal for the Dartmouth summer research project on artificial intelligence. http://www-formal.stanford.edu/jmc/history/dartmouth/dartmouth.html. Accessed October 2006
2. Russell, S.: Human Compatible: Artificial Intelligence and the Problem of Control. Viking, New York (2019)
3. Russell, S., Norvig, P.: Artificial Intelligence: A Modern Approach, 4th edn. Prentice Hall, Upper Saddle River (2020)
4. Görz, G., Schmid, U., Braun, T.: Handbuch der künstlichen Intelligenz, 5th ed. De Gruyter, Berlin, p. 976 (2020). (in d)
5. Pearl, J., Mackenzie, D.: The Book of Why: The New Science of Cause and Effect. Basic Books, New York (2018)
6. Müller, V.C.: Ethics of artificial intelligence and robotics. In: Zalta, E.N. (ed.) Stanford Encyclopedia of Philosophy, vol. Summer 2020, pp. 1–70. Palo Alto: CSLI, Stanford University (2020)

7. Eurobarometer: Public attitudes towards robots, vol. 382. http://ec.europa.eu/public_opinion/archives/ebs/ebs_382_sum_en.pdf
8. Eurobarometer: Autonomous systems, vol. 427. https://ec.europa.eu/commfrontoffice/publicopinion/archives/ebs/ebs_427_en.pdf
9. Müller, V.C.: History of Digital Ethics. In: Véliz, C. (ed.) Oxford Handbook of Digital Ethics. Oxford University Press, Oxford. forthcoming
10. Coeckelbergh, M.: AI Ethics. MIT Press, Cambridge, Mass (2020)
11. Gibert, M.: Faire la moral aux robots: Une introduction à l'éthique des algorithmes. Montréal: Atelier 10 (2020)
12. Gordon, J.-S., Nyholm, S.: Ethics of Artificial Intelligence, Internet Encyclopedia of Philosophy. https://iep.utm.edu/ethic-ai/
13. Floridi, L., Saunders, J.W.: On the morality of artificial agents. Minds Mach. **14**, 349–379 (2004)
14. Moor, J.H.: The nature, importance, and difficulty of machine ethics. IEEE Intell. Syst. **21**(4), 18–21 (2006). https://www.researchgate.net/publication/220629129_The_Nature_Importance_and_Difficulty_of_Machine_Ethics
15. Anderson, M., Anderson, S.L. (eds.): Machine Ethics. Cambridge University Press, Cambridge (2011)
16. Wallach, W., Asaro, P.M. (eds.): Machine Ethics and Robot Ethics. Routledge, London (2017)
17. Müller, V.C.: Is it time for robot rights? Moral status in artificial entities. Ethics Inf. Technol. 1–9, forthcoming. https://www.springer.com/journal/10676
18. European Commission: White paper on artificial intelligence: A European approach to excellence and trust, vol. COM (2020) 65, no. 19.2.2020, pp. 1–27 (2020). https://ec.europa.eu/info/publications/white-paper-artificial-intelligence-european-approach-excellence-and-trust_en
19. Hagendorff, T.: The ethics of AI ethics: an evaluation of guidelines. Mind. Mach. **30**(1), 99–120 (2020).https://doi.org/10.1007/s11023-020-09517-8
20. Jobin, A., Ienca, M., Vayena, E.: The global landscape of AI ethics guidelines. Nat. Mach. Intell. **1**(9), 389–399 (2019). https://doi.org/10.1038/s42256-019-0088-2
21. AI HLEG: High-level expert group on artificial intelligence: ethics guidelines for trustworthy AI, European Commission, pp. 1–37 08 April 2017. https://ec.europa.eu/digital-single-market/en/high-level-expert-group-artificial-intelligence
22. Mittelstadt, B.: Principles alone cannot guarantee ethical AI. Nat. Mach. Intell. **1**(11), 501–507 (2019). https://doi.org/10.1038/s42256-019-0114-4
23. Van de Poel, I.R., Royakkers, L.M.: Ethics, Technology and Engineering. Wiley-Blackwell, Oxford (2011)
24. Mittelstadt, B.D., Floridi, L.: The ethics of big data: current and foreseeable issues in biomedical contexts. Sci. Eng. Ethics **22**(2), 303–341 (2016). https://doi.org/10.1007/s11948-015-9652-2
25. Floridi, L., Taddeo, M.: What is data ethics? Phil. Trans. R. Soc. A **374**(2083) (2016)
26. Morley, J., Elhalal, A., Garcia, F., Kinsey, L., Mokander, J., Floridi, L.: Ethics as a service: a pragmatic operationalisation of AI Ethics, 11 Feburary 2021. arXiv:2102.09364v1. https://arxiv.org/abs/2102.09364
27. Stix, C.: Actionable principles for artificial intelligence policy: three pathways. Sci. Eng. Ethics **27**(1) 1–17 (2021). https://doi.org/10.1007/s11948-020-00277-3
28. HLEG: Assessment List for Trustworthy AI (ALTAI) for self- assessment no. 17 June 2020. https://futurium.ec.europa.eu/en/european-ai-alliance/pages/altai-assessment-list-trustworthy-artificial-intelligence

29. Floridi, L., Cowls, J., King, T.C., Taddeo, M.: How to design AI for social good: seven essential factors. Sci. Eng. Ethics **26**(3), 1771–1796 (2020). https://doi.org/10.1007/s11948-020-00213-5
30. Wahlster, W., Winterhalter, C.: Deutsche Normungsroadmap Künstliche Intelligenz, p. 232. DIN/DKE, Berlin (2020)

Check for updates

Ethical Issues Relating to the Environmental Impact of Social Robots

Fiachra O'Brolcháin[1,2](✉)

[1] Institute of Ethics, School of Theology, Philosophy and Music, Dublin City University, All Hallows, Dublin, Ireland
fiachra.obrolchain@dcu.ie
[2] DCU Centre for Climate and Society, Dublin City University, Dublin, Ireland

Abstract. This extended abstract focuses on the environmental impact of robotics. While robots have already proved extremely beneficial to humanity and new developments promise continued benefits, particularly in the area of social robotics, it would be unwise to neglect the environmental impact of this revolution. The environmental impact of robotics can be more or less sustainable – the outcome will depend on the choices made now. As such, these are ethical issues – the decisions taken now by policy makers, regulators, developers, and others will impact the lives of current and future generations. These choices are not neutral, purely technical choices, but choices with long-term consequences. As such they are ethical choices and require normative analysis. This piece sketches some of the ethical issues that must be considered.

1 Introduction

Environmental externalities of technological progress can no longer be ignored in good conscience.

The coming decades are likely to see a vast increase in the availability and use of social robotics. This will include, but is not limited to, the use of exoskeletons, prosthetics, humanoid robots, and collaborative robots. Robots may or may not include artificial intelligence (AI), but all robots are distinguished by being physical artefacts that can move to a certain extent and are capable of interacting with the physical world. Advances in robotics mean they have the potential to be truly revolutionary. As such, they provoke myriad ethical questions.

Discussing the ethical issues associated with the environmental impact of robotics is a complex issue -philosophical methods of ethical analysis are utilised here. In the following section, the background context will be sketched. The discussion will outline the main areas of ethical concern.

2 Background

While the environmental impact of robotics might not seem the obvious starting point for an ethical discussion of robotics, the environmental context of any human activity – particularly one that might transform society in a deep and pervasive way (such as

M. A. Grau Ruiz (Ed.): INBOTS 2021, BIOSYSROB 30, pp. 10–13, 2022.
https://doi.org/10.1007/978-3-031-04305-5_2

robotics) – cannot be ignored. The IPCC's Summary for Policy Makers of the Physical Science Basis of the Sixth Assessment Report is unequivocal: "The scale of recent changes across the climate system as a whole and the present state of many aspects of the climate system are unprecedented over many centuries to many thousands of years" [1]. We are moving towards a planetary climate system vastly different to the climate system in which human civilization developed. As UN Secretary General, António Guterres, said, this is a "code red" for humanity [2].

The development of social robotics takes place against this background. Currently it is estimated that around 500, 000 new industrial robots come into service annually [3]. Social robots are not yet at that scale though there is the possibility that social robotics might become pervasive in society.

Whilst robots might have benefits from an environmental perspective, they will certainly bring environmental challenges with them. For instance, they will require raw materials. This will need to be extracted and transported. These supply chains will have a carbon footprint. Robots will need to be assembled and will require power to function. Robots, once obsolete, will need to be disposed of or recycled. The rapid pace of the development of the Internet, mobile phones and other communications devices over the past few decade provides a useful analogy – the technologies were pursued and celebrated, various ethical concerns related to their social and political impact then arose (e.g. privacy concerns, filter bubbles). The environmental impact of these technologies was given little attention, but is now known to be significant: digital technologies contribute up to 3.7% of global emissions [4].

The widespread manufacture, development and use of robotics and the infrastructure used to maintain them will (under current industrial practices) increase humanity's collective carbon footprint and make it more difficult to avoid the more severe and catastrophic impacts of climate change. The development of robots, and of a society in which social robots are commonplace or pervasive, will not be without impact. This will take place at the level of individual robots and at a collective level.

3 Discussion

Given this background, it is suggested that the environmental impact of robotics is of immense ethical significance and ethics should play a significant role in the development of the field. The simplest questions relate to whether the development of robots is truly justified at all – is this something on which valuable time, energy and resources should be spent? Whilst this is possibly of philosophical interest, it is of limited practical use as robotics development is already well under way. Whilst certain ethical issues have been of interest to robotics developers, environmental issues have not been amongst them. Instead, the focus has been on issues such as privacy and responsibility.

High-level ethical principles such as those found in deontology, consequentialism, or virtue theory can play a role, though they may be of limited practical use to those either developing robots or designing policy or regulation in relation to them. Of more immediate practical use are the Sustainable Development Goals (SDGs). Under the title "Transforming our World: The 2030 Agenda for Sustainable Development" [5], the 193 members of the United Nations ratified the Sustainable Development Goals (SDGS) in September 2015. These goals can be seen to be normative in that they are held in common and set goals towards which all citizens and signatory nations should aspire if a sustainable future is to be feasible. There are 17 SDGs and thus there exists risk of conflict and of dilution of the environmental impact. For instance, Goal 9: Industry, Innovation and Infrastructure see allow unscrupulous developers to pursue profitable but environmentally deleterious practices under the flag of sustainable development. As such, it is essential the SDGs, if they are to be used properly as a normative guide, be taken as a whole and viewed holistically. Taken as such the SDGs will provide a baseline against which the development of social robotics can be assessed from an environmental perspective.

Beyond this, there are a number of specific areas that will need to be explored when analysing the environmental issues arising from robotics. Questions of justice and moral responsibility will arise in relation to the distribution of the environmental costs and benefits associated with the use (and eventual disposal) of social robotics, i.e. will any environmental burdens (waste, pollution) be borne by those benefiting from the use of the robots and will the benefits of these novel technologies be available to all?

In these areas, what will be required are more specific detailed arguments that can be readily translated into regulatory instruments. To this end, a green agenda for robotics is urgently required. The environmental implications of the entire lifecycle of a robot (including extraction, transportation, manufacturing, and intended use, and disposal) need to be addressed in guidelines for robotics developers and in legal and financial regulation. Such an agenda is likely to be based on one of the traditional ethical frameworks (as mentioned above) but should be cognisant of new research areas focused on biocentric and ecocentric ethics.

The environmental ethics framework of social robotics should resemble the Responsible Research and Innovation (RRI) initiative that the European Union has focused on recently. Whilst RRI initiatives placed some emphasis on the environmental concerns, these concerns should be front and centre of new initiatives governing technological development. Basing technological transformation of society was built on a technological revolution that reduced our collective impact on the environment, the benefits would be immense and long-lasting. Given the scale of the climate problems currently facing us, it seems obvious there is a collective responsibility on us to pursue such transformations.

Designers, robotics developers, policymakers and legislators need increased awareness of the ethical implications of the environmental impact of social robotics. This requires further philosophical research into what these implications are the normative guidelines that should underpin the future development of social robots. This can in turn inform legal experts as they look to create the frameworks that will govern the future development of social robots. This is in-line with general shifts in technology research and engineering away from "technology only" towards transdisciplinary collaboration.

4 Conclusion

Ethical research and analysis relating to the environment should play a central role in the development of social robotics. The majority of the world's nations have already signed up to the UN's SDGs and accepted general obligations to pursue sustainable development. The IPCC reports are unambiguous regarding the severity of the current climate situation. The development of social robots is taking place against this background. In this context, policymakers and legislations have an obligation to ensure that social robots do not impede efforts to ameliorate the worst impacts of climate change. Existing legislative and financial tools should be utilised to ensure that the nascent social robots industry consists of environmentally sustainable practices; and more explicitly-environmentally focused frameworks should be drawn up. This will require further philosophical analysis and, in conjunction with the social robotics industry, environmentally focused ethics guidelines should be developed and integrated into the social robots industry.

Acknowledgement. This research was carried out under the INBOTS project, which has received funding from the European Union's Horizon 2020 research and innovation program under Grant agreement No 780073.

References

1. I.P. on C.C. IPCC: Summary for Policy Makers in Climate Change 2021: The Physical Science Basis. Contribution of Working Group I to the Sixth Assessment Report of the Intergovernmental Panel on Climate Change (2021)
2. U. N. UN: Guterres: The IPCC Report is a code red for humanity, Code Red (2021). https://unric.org/en/guterres-the-ipcc-report-is-a-code-red-for-humanity/. Accessed 22 Aug 2021
3. Müller, V.C.: Ethics of artificial intelligence and robotics. In: Zalta, E.N. (ed.) The Stanford Encyclopedia of Philosophy, Summer 2020, Metaphysics Research Lab, Stanford University (2020). https://plato.stanford.edu/archives/sum2020/entries/ethics-ai/. Accessed 12 June 2020
4. Griffiths, S.: Why your internet habits are not as clean as you think (2020). https://www.bbc.com/future/article/20200305-why-your-internet-habits-are-not-as-clean-as-you-think. Accessed 23 Aug 2021
5. UN General Assembly: Transforming Our World: the 2030 Agenda for Sustainable Development. United Nations (2015). https://sustainabledevelopment.un.org/sdgs. Accessed 12 Mar 2020

Human Vulnerability and Inclusive Robotics

Belén Liedo[1(✉)] and Jon Rueda[2]

[1] Institute of Philosophy, Spanish National Research Council (CSIC), Madrid, Spain
belen.liedo@cchs.csic.es
[2] University of Granada (UGR), Granada, Spain
ruetxe@ugr.es

Abstract. Vulnerability is one of the key ethical concepts on the evaluation of interactive robotics and other potentially disruptive technologies. Recently, vulnerability has been a central concept in several ethical reflections and, also, guidelines and ethical standards. A correct acknowledgment of vulnerability is essential for responsible research and innovation on Robotics and Artificial Intelligence. Here, we argue for specific attention to vulnerability when evaluating the ethical adequacy of potentially disruptive technologies, particularly in the case of interactive robotics. First, we propose a version of vulnerability inspired by the work of Florencia Luna and Mark Coeckelbergh. Then, we briefly analyse some debates about vulnerability that has been prominent among controversies around transhumanism and human enhancement. Finally, we propose three ethical standards for taking vulnerability into account in ethically responsible technological research.

1 Introduction

In this paper, we will advocate for special attention to the issue of vulnerability in ethical assessment of interactive robotics. Human vulnerability is a complex, relevant concept in contemporary applied ethics. We argue that this is a useful conceptual tool for building inclusive interactive robotics. In doing so, we follow a tripartite structure. First, we develop an account of vulnerability illuminated by recent, most influential versions of the concept. Then, this concept is applied to the field of disruptive technologies, through the reflection on contemporary debates among transhumanist proposals. Finally, we propose three ethical standards suitable for the analysis of interactive robotics.

2 Human Vulnerability

VULNERABILITY has recently been claimed as a defining category of human beings. It can be understood in two different senses: (a) as an existential, universal, ontological category, and (b) as a situational, political, and contingent category. Both ideas retain something salient of the human condition. On the one hand, it is undeniable that human

This paper benefits from Grant Agreement 780073-INBOTS, Spanish Ministry of Universities grant FPU19/06027, and INPhINIT Retaining Fellowship of the La Caixa Foundation LCF/BQ/DR20/11790005.

M. A. Grau Ruiz (Ed.): INBOTS 2021, BIOSYSROB 30, pp. 14–18, 2022.
https://doi.org/10.1007/978-3-031-04305-5_3

beings are frail and mortal, and that this is a universal feature. Ontological vulnerability refers to this idea, that has its precedents in occidental philosophy tradition. On the other hand, one must also recognize that harm and peril are socially determined issues. Social or situational vulnerability is the concept that refers to this meaning. Severity of the harms one is exposed to rely mostly on the social and political structure one is born into. Vulnerability is thus a universal condition, but its arising factors are unequally distributed within human species because of diverse social and political situations.

The category of vulnerability has been present in different ethical guidelines and frameworks in the last decades [1]. Sometimes, it appears in the form of "vulnerable groups". Vulnerable groups may be defined as "those incapables of protecting their own interests" [2]. It has been criticized that this concept can lead to stigmatization and marginalization for those groups. For avoiding this problem, we follow here the proposal made by Florencia Luna [3]. As stated by Luna, people are embedded in variable vulnerability situations. These situations work as "layers": they can overlap, change, and interact. The specific situation of vulnerability derived, for example, for being a woman, depends on contingent conditions. Being a woman does not make one necessarily vulnerable, as a label; better, it is a situation that can create some specific vulnerabilities that are susceptible to be tackled and managed through adequate ethical and social evaluations.

3 Vulnerability and Technology

Social vulnerability refers to the conditions of a certain situation, understood as culturally determined. That means that social vulnerability is contingent and susceptible to change. Social vulnerability, then, is a way of critically evaluating power relations among different phenomena.

As a human issue, technology has always something to do with vulnerability. Following Mark Coeckelbergh [4, 5], technology is one of the many tools that humans use to manage vulnerability. Technological changes in societies usually carry important changes in the way of living of individuals. Mostly, technology helps us to overcome some of our natural imperfections, but it can also create new kinds of vulnerabilities.

Ontological vulnerability is a universal condition. In a sense, this statement is an undeniable truth. However, this acknowledgment should not fail to understand the complexity of vulnerability. Some authors have claimed that, since vulnerability define us as human, some technological aspirations should be tempered and stop seeking invulnerability [6]. Here, we argue that this quest for invulnerability is neither possible nor desirable, but we also defend that reducing vulnerabilities is what we do as human species: aiming to lessen harm and peril in life.

In consequence, there are two main arguments related to technology and vulnerability:

(a) Vulnerability will not disappear by the action of technology.
(b) Aiming to reduce vulnerability does not mean aiming to eradicate vulnerability, which is not a desirable objective in itself.

In the first place, it must be noted that not even the most prominent figures of the transhumanist movement defend something such as invulnerability [7]. In the ontological sense, vulnerability cannot be avoided. We will always be vulnerable to harm since we are inserted into a world. Even if we were not inserted (physically) into a world, we will be still vulnerable to death. So, utopian tendencies that can be found sometimes in remarkably radical technological projects need to be tempered [8].

When talking about diminishing vulnerabilities, we should not forget the social variant of the concept. Harm and peril are unequally distributed among societies and groups of people. Indeed, not everyone suffers the same degree of exposure to some universal threats to human life, for example, extreme weather conditions. In this vein, all technological advances that seek to overcome these perils are *a priori* welcomed; but one should always keep in mind the side-effects that each technology may create. Overcoming vulnerability for some may mean exacerbating vulnerabilities for some others.

In the second place, we accept that vulnerability is somehow defining of humanity, but not only in its negative sense. As stated by Nussbaum [9], vulnerability is the other side of the coin of our open-ended and creative condition. We can suffer because we are responsive to the world; we are never fully adapted to the environment and thus we change the medium [10]. Our vulnerability, in a sense, is what elicits our freedom and development.

Furthermore, we must acknowledge the moral value of caring relationships, which are intrinsically related to vulnerability. We are vulnerable, therefore we need to be cared for by others. The caring network we establish with other people has a moral value that we should want to preserve.

4 Technology-Friendly, Non-essentialist Account of Vulnerability

Robotics and Artificial Intelligence (AI) may be considered as socially disruptive technologies [11], and therefore they deserve detailed ethical scrutiny. However, it is not clear whether they should be considered essentially different from other previous technologies when it comes to vulnerability evaluations. If we consider human beings as technological beings, we can understand new technologies as a continuity of other previous technological interventions. In this regard, interactive robotics is not understood as an existential threat to the human species; rather, it is understood as a new way to cope with the world. With the creation and dissemination of interactive robotics, we should not focus on ontological vulnerability. We should be concerned about its impact on social vulnerabilities, aiming to maximize the impact that robotics can have in reducing harm and inequalities among different populations, and avoiding increasing occurrent vulnerabilities [12].

Transhumanism is the movement that has made the most prominent claims about future potential of these and other technologies for changing the future of humanity; concretely, for even transforming human species into a new species. Even if some have claimed that these aspirations will end vulnerability, we have already stated that this is impossible. We propose that there are two main versions of transhumanism and that there is a possibility for a moderate, vulnerability-friendly one. In that last case, it would be possible to reconcile the aim for technological advance with the acknowledgment and management of vulnerability.

Some contemporary robotics and AI are increasingly invasive, intended to co-operate closely with human beings. Interactive robotics will impact situations of vulnerability in which people are embedded. Some of these foreseeable variations may not change human nature as a whole, as stated in the most radical transhumanist proposals, but may change the situation experienced by a group of subjects regarding vulnerability.

If we define human beings as technological beings, then technology is not something external to our nature, but it is embedded in our very condition as embodied and creative beings. So, consequently, technological vulnerability is a particularly important matter.

We recommend that vulnerability is included as an evaluation factor for ethical guidelines in inclusive robotics. In our view, acknowledging vulnerability implies at least three ethical standards:

- *Precaution.* Vulnerability transformations through technology may lead to unpredicted situations. Precaution will be needed to avoid as much as possible the negative, unforeseen side-effects of robotics.
- *Politics of vulnerability.* Social vulnerability is unfairly distributed across populations. Therefore, vulnerability transformations due to technology would impact differently depending on the particular situations. Ethical evaluation needs to be capable of anticipating differential impacts in all the stakeholders. For this evaluation, one needs to take into account that unequally distributed conditions of vulnerability can lead to social marginalisation or some groups, and that special attention needs to be paid to these marginalized realities to include them in ethical evaluation.
- *Social responsibility in design.* As stated, even if all humans are ontologically vulnerable, conditions of arising of certain vulnerabilities have political and social roots. As such, these vulnerabilities are contingent and variable. Responsible research should be committed to (i) not worsen current situations of vulnerability for certain groups of people, and (ii) progressing towards a better and more effective managing of vulnerability, reducing harm and peril as much as possible.

Ethically responsible technology should consider the variety and variability of vulnerability situations across populations. For doing so, it is recommended to keep in mind the double nature of vulnerability. On the one hand, it is a universal condition that makes us all dependent on one another, and that makes possible a universalizable solidarity. On the other hand, social determinants of vulnerability need to be tackled by technology, with the general objective of reducing harm and inequalities.

5 Conclusion

As stated, vulnerability is a defining feature of the human condition, but so it is the aim for reducing this exposure to harm. Any technological endeavour that aspires to be legitimate needs to acknowledge the role vulnerability plays in human societies. Particularly, naïve aspirations for invulnerability should be discarded. For what it refers to ontological vulnerability, it will probably continue to be a radical condition for human life. On the other hand, social vulnerability is dependent on power relations and political factors. Technology should help to diminish inequalities related to vulnerability contexts and prevent creating novel unexpected situations of vulnerability by failing to correctly evaluate the consequences of each innovation.

Ethical evaluation regarding technology needs to take specific account of vulnerability. In order to do so, we have proposed three ethical standards that aspire to be combinable and complementary with other general guidelines: precaution, politics of vulnerability, and social responsibility in design.

Interactive robotics has the potential of successfully managing some situations of vulnerability experienced by people all over the world. But this objective will not be immediate nor evident. A correct evaluation of vulnerability transformations for a concrete technology will be crucial if we want to create truly inclusive interactive robotics.

References

1. Liedo, B.: Vulnerabilidad. EUNOMÍA Revista en Cultura de la Legalidad **20**, 242–257 (2021)
2. Council for International Organizations of Medical Sciences (CIOMS). (2002 [1998]): International Ethical Guidelines for Biomedical Research Involving Human Subjects. https://cioms.ch/publications/product/international-ethical-guidelines-for-biomedical-research-involving-human-subjects-2/
3. Luna, F.: Identifying and evaluating layers of vulnerability–a way forward. Dev. World Bioeth. **19**, 86–95 (2019)
4. Coeckelbergh, M.: Vulnerable cyborgs: learning to live with our dragons. J. Evol. Technol. **22**(1)1–9 (2011)
5. Lemmens, P.: Coeckelbergh, Mark: Human Being @ risk: Enhancement, Technology, and the Evaluation of Vulnerability Transformations, vol. 12. Springer Science & Business Media, Heidelberg (2013). https://doi.org/10.1007/s10746-014-9308-2
6. Hauskeller, M.: Ephemeroi–human vulnerability, transhumanism, and the meaning of life. Scientia et Fides **7**(2), 9–21 (2019). https://doi.org/10.12775/SetF.2019.013
7. Bostrom, N.: Why I want to be a posthuman when I grow up. In: Medical Enhancement and Posthumanity, pp. 107–136. Bert Gordijn & Ruth Chadwick, Springer, Dordrecht (2008)
8. Liedo Fernandez, B., Rueda, J.: In defense of posthuman vulnerability. Scientia et Fides **9**(1), 215–239 (2021). https://doi.org/10.12775/SetF.2021.008
9. Nussbaum, M.C.: The Fragility of Goodness. Luck and Ethics in Greek Tragedy and Philosophy. Cambridge University Press, Cambridge (1986)
10. Gasset, J.O.Y.: Ensimismamiento y alteración. Meditación de la técnica y otros ensayos. Alianza (2014)
11. Rueda, J., Ausín, T.: Interview with Philip Brey. DILEMATA **31**(13), 133–137 (2021). https://www.dilemata.net/revista/index.php/dilemata/article/view/412000418
12. Mackenzie, C., et al.: Vulnerability. New Essays in Ethics and Feminist Philosophy. Oxford University Press, Oxford (2014)

Robot-Human Gaze Behaviour: The Role of Eye Contact and Eye-Gaze Patterns in Human-Robot Interaction (HRI)

Aníbal Monasterio Astobiza[1](✉) and Mario Toboso[2]

[1] Universidad de Granada, Granada, Spain
anibalmastobiza@gmail.com, amastobiza@ugr.es
[2] IFS-CSIC, Madrid, Spain

Abstract. We want to explore the idea that for our naturally evolved social cognition to be applied to artificial agents (e.g. social robots) a particular configuration of the physical characteristics of the robot (e.g. physical embodiment) is needed. For example, key social features or a topography of elements similar to a human face where the eyes are essential.

1 Introduction

HUMAN beings are social animals [1]. Thanks to our ability to ascribe mental states and predict behaviour (e.g. social cognition) we are able to act together (e.g. joint action). Social cognition gives us the opportunity -given certain information about the mind and behaviour of others- to make the best decision. The mere possibility of acting together or joint action shows us that sometimes the best decision is the one we make collectively instead of on our own. So, in this sense, social cognition and joint action are two intertwined processes. There is no consensus on a definition of social cognition because it has a number of different components and a diverse array of new disciplines take it as an object of study. So, for example, in primatology, developmental and comparative psychology, social cognition is called "theory of mind" [2, 3], in philosophy has received the name "intentional stance" [4] or "mindreading" and with the work of pioneers like Leslie Brothers, John Cacioppo, Ralph Adolphs and others in the union of cognitive neuroscience and social psychology, leading to the creation of the new subdisciplines of social cognitive neuroscience within social neuroscience, it is used the term "neural hermeneutics" [5] to refer to social cognition. According to Chris Frith [6], social cognition is just cognitive psychology applied to social psychology. Cognitive psychology is a young subject that started in the late twentieth century replacing behaviourism in the study of people. The cognitive in cognitive psychology basically refers to the fact of using computers as a metaphor. Hence, the scientific community began to talk about memory as storage, perception as information processing and representations.

This work was supported by the projects INBOTS (Ref. 780073) and EXTEND (Ref. 779982), EthAI+3 (PID2019-104943RB-100), and the project "Capacitismo" (FFI2017-88787-R) of the Retos Programme of the Spanish Government R+D+I Plan.

M. A. Grau Ruiz (Ed.): INBOTS 2021, BIOSYSROB 30, pp. 19–24, 2022.
https://doi.org/10.1007/978-3-031-04305-5_4

Therefore, social cognition is just the application of the ideas of cognitive psychology to social psychology. If cognition is no more than the umbrella term to refer to basic psychological processes such as perception, attention, memory and action planning… social cognition is the study of how all these processes take place in social contexts when two or more people interact together. In other words, social cognition can be seen as the human capacity to navigate the social world. In this social world, there are other people with whom we interact on a daily basis. But we also interact with other non-human agents. A subset of these non-human agents with which we interact comprised other animals (e.g. dogs, cats…). However, progressively artificial agents are also part of our social world. The ubiquity of Artificial Intelligence (AI) and the growing presence of humanoid robots (AI and robotics are two sides of the same coin because robotics is the tangible and mechanical embeddedness of AI) challenges our naturally evolved social cognition to deal with people and other non- human animals. We define social robots as: artificial systems with sensors, actuators and AI to process information in order to enact actions whose function is to communicate with and respond appropriately to its environment including humans. Artificial agents pose an important test to social cognition: How do we understand artificial agents? and What are the consequences of attributing mental states to artificial agents? In this short paper, we want to explore the idea that for our naturally evolved social cognition to be applied to artificial agents (e.g. social robots), a particular configuration of the physical characteristics of the robot (e.g. physical embodiment) has to be present. For example, key social features or a topography of elements similar to a human face where the eyes are essential. We are agnostic about the true nature of social cognition and its function, whether or not social cognition aims to gain knowledge about the inner states of people (or even artificial agents such as social robots) or whether social cognition just simply works as a mechanistic-driven explanation tool to make sense of behaviour and manage the social image we project to others. Nonetheless, we believe social cognition can be deployed in the absence of any linguistic behaviour or verbal input, because it is sufficient the presence of non-verbal cues such as facial expressions or non-verbal behaviour in general.

2 Joint Attention and Joint Action with Social Robots

There is close relationship between joint attention, joint action and social cognition or as we prefer to say in this context: "mind sharing". Joint attention occurs when two or more individuals orient their attention towards a third object, event or other individual and usually occurs because one individual initiates the process of establishing eye contact with some other and directs him/her to a third object, event or another individual [7]. However, for a joint action to occur no matter how simple the joint action is, joint attention is not enough. Joint action, shared cooperative activity, has several essential characteristics [8]:

- Mutual responsiveness
- Commitment to joint activity
- Commitment to mutual support

Batman's account of joint action has been the subject of some criticisms due to the fact that seems to require sophisticated cognitive capacities and a higher level development of social cognition or at least shared intentionality. Hence, it becomes difficult to see how this characterization of joint action enables artificial agents or even infants to engage in joint action. However, we believe that Batman's analytical distinctions support our idea that eyes are essential. Because at the level of mutual responsiveness, without the intuitive recognition of some eye-like stimuli that our human social cognition depends on, it would be impossible any mutual responsiveness.

Because at the level of commitment to joint activity, without eye-gaze behaviour (e.g. fixated gaze on an object/event/person and shift in gaze from an object/event/person to another environmental location), it would be impossible any commitment to joint activity or at least its recognition. We recognize that interaction is a complex phenomena rooted in a multimodal sensory-motor processes but some minimal humanoid/anthropomorphic appearance (physical structure with eye-shaped like stimuli) creates a real sense of interaction. To have a genuine interaction between human beings and artificial agents, such as social robots, we believe that it is necessary to have a specific physical topography of elements, where eyes or stimuli similar to the eyes are essential. Eye-shaped like stimuli seems to be a critical part when engaging with a social robot. Of course, robot eyes could be of different forms and shapes and eye-gaze behavior do not need to correspond with a true visual perception system. But the role of eye-like stimuli smoothes the human-robot interaction.

3 Feature Engineering for HRI: The Necessity of the Eyes

To encapsulate the idea that social robots engage in a shared cooperative task (joint action) with humans, we propose the following thesis:

(Thesis) The design-perception equivalence: design implies perception in HRI (Human-Robot Interaction).

- Eye-gaze behavior improves interaction
- Social robots are designed for HRI (Human-Robot Interaction)
- Therefore, the perception of certain appearance (eyes) in social robots serves the goal of HRI

The physical structure of a robot is an extremely important aspect in robot design and perception as well.

4 Conclusion

We believe that the basic physical structure of a robot and its design with eye-shaped stimuli is necessary to give rise to an experience of genuine interaction between human beings and artificial agents such as social robots. The basic geometry and structure of a social robot needs to have a head and eye-shaped like stimuli. We believe that to have a genuine interaction between human beings and artificial agents, such as social robots, it is necessary to have a specific physical topography of elements, where eyes or stimuli similar to the eyes are essential. Eye-shaped like stimuli seems to be a critical part when engaging with a social robot. Of course, robot eyes could be of different forms and shapes and eye-gaze behavior do not need to correspond with a true visual perception system. But the role of eye-like stimuli smoothes the human-robot interaction. The physical structure of a robot is an extremely important aspect in robot design and perception as well. The chassis, its geometry, or even how to build the robot frame in terms of materials, degrees of freedom etc. are areas of research in themselves with dedicated specialists. In its design many people collaborate, including psychologists and philosophers. The basic geometry of a social robot needs to have a head and eye-shaped like stimuli. In the previous sections, we saw different complementary models that characterize joint attention, joint action and its relationship with social cognition. Our action perception system is not only selectively activated by the perception of biological movement. This action perception system is quite versatile and can be activated by biological movement and by the movement of an artificial agent. But there are certain structures in our social brain that respond preferentially to human-shaped forms, such as the eyes, that allow for more effective interaction.

Specifically, gaze-induced human-robot interaction, via joint attention, is a mechanism that allows two parties to attend to a third object or event in the environment which has a very important role to play in facilitating cooperation and achieving common tasks. In a world where we increasingly share space with robots and artificial systems, if they are to be truly social robots, they need to respond to social signals such as those provided by eyes. The human socio-cognitive system is flexible, but has evolved to respond selectively to relevant stimuli that carry social signals. If we really want to apply robots in society and we want them to be social and inclusive robots that take into account people's needs and interests, we have to recognize that the design of these artificial systems must be based on the mechanisms of our social cognition that have been developed to interact with other people, not artificial agents. Therefore, the design of our social robots must correspond to the stimuli that induce a response in our action perception, attention, and higher-order social cognition mechanisms. Only in this way will we be able to facilitate a natural interaction between a robot and a human being. If the robot has eye-shaped stimuli, it will facilitate our interaction with it. And if it is possible, if this artificial agent is able to follow our gaze, we will be able to perform tasks together.

As in the coming decades more and more robots will be applied in society in multiple sectors and for multiple tasks, we can only ensure that there will be meaningful interactions with robots if we design robots so that our social brains respond as they do when we interact with other people. This will mean changes in the representation and external appearance of robots that will allow appropriate anthropomorphic inferences. To integrate robots into human environments, we need to design them for genuine and effective natural interaction. We can only do this by taking into account robot-human gaze behaviour and the fundamental importance of the role of eye contact and eye-gaze patterns in human-robot interaction. Only by applying knowledge of the processes and mechanisms behind human social cognition and its potential relationship to other faculties [9] can we better understand how humans understand robot stimuli.

Let us conclude this brief review of the application of social cognition to human-robot interaction and in particular the recognition of the importance of robot-human gaze behaviour, with a list of the various types of eye gaze using the terminology that appears in the research literature:

- Mutual gaze: It is the gaze that is directed from one agent to the eyes or face of another, and vice versa. The gaze directed to the face without reciprocity is not a mutual gaze.
- Referential gaze or deictic gaze: It is the gaze aiming at attracting visual attention of a communication partner to a referred entity in the environment
- Gaze aversions: It is the discontinuation or intentional avoidance of eye contact

Eye gaze plays a fundamental role in human communication and interaction, and of course it also plays a fundamental role in robot-Human (gaze) behaviour. People to refer to objects in their environment look at them before naming them or even manipulating them. People are able to determine with great precision the direction of the gaze and the target of the speaker's gaze and use these social cues to predict what they are going to say. Eye gaze also correlates with the content of what people say. Members of a group performing a joint task coordinate their eye gazes as a non-intrusive communication channel. When they are interacting with each other, they use their eye gaze to know in advance what they may be referring to. On the other hand, when there are ambiguous terms in speech, the eye gaze is a mechanism to disambiguate and fix to which object in the environment, person or event is being alluded. When people manipulate objects, the gaze anticipates the intentional action and often anticipates the physical action, such as moving the hand. The shift of gaze from one object to another anticipates the next movement (e.g. kinematic aspect). These shifts in eye gaze often correspond to the beginning of a new action (or kinematic aspect) in the manipulation of an object. This implies that objects that are not part of or related to the task at hand are rarely fixated. Of course, eye gaze also signals visibility or the opportunity to interact. Staring at someone is not the same as not looking at them, since looking at them means availability for potential communication.

We are particularly tuned to attend to the eye gaze because of the importance of what it signals (e.g. the direction of people's attention, their states of mind etc.) The eyes are sensory organs that have among their functions the formation of visual categories, but this is not their only function. In humans, the sclera, the white part of the eyeball, is highly exposed and considerably more elongated than in other primate species. In 88 primate species studied, Kobayashi and Koshima found no white sclerae, these are camouflaged with the coloring of the skin or facial hair and serve as an adaptation to avoid discrimination of gaze orientation because in many cases direct eye contact provokes attacks and conflicts [10]. However, this is not the case for the morphology of the human eye. According to these researchers, the human eye is adaptively designed not only as a sensory organ but also as a communication device. For example, the human brain has evolved highly sensitive neural circuits, anatomically located in the temporal cortex, that respond to the movement and direction of gaze [11]. This region of the brain, known as the superior temporal sulcus (STS) is an essential part of the "social brain".

In this sense, the eyes are not only sensory organs but also have other social functions as communication devices. The social functions of gaze fixation on another person's eyes (or even robots's eyes) are:

- Turn-taking control, an intermittent gaze or eye contact signals to the listener that it is his or her turn to speak.
- Intimacy, a long-lasting stare is a sign of acceptance in the case of romantic partners.
- Control and submission, a steady gaze is a sign of control of the situation and is used to intimidate, while an eye gaze that is withdrawn indicates submission. The existence of a relationship between gaze ethology and social hierarchy is considered to exist.
- Cooperation and communication, an eye contact may indicate that one expects the contribution of the other and also through the gaze certain informational messages can be transmitted.

Learning from these aspects of social cognition and the social brain and how gaze is fundamental to human interaction allows us to understand how to design better social robots for the interaction with people. Eye gaze and specifically robot-human gaze behaviour supports and augments social behaviours between robots and humans. Social robots could take advantage of eye gaze behaviour to convey intentions, mental states, send social signals, in a natural and human like way. Taking advantage of the eye gaze and robot-human gaze behaviour would improve the design of social robots and thus allow us to have real and genuine interactions with robots. The physical embodiment of social robots needs eyes. Eye gaze has a tremendous impact on human-robot interactions. Robots with the right physical embodiment, expressive eyes, interact better with humans.

In the context of HRI, it is impossible to promote a successful interaction, let alone intersubjectivity or affective behavior if there are not affordances to elicit these responses. For us, the eyes have it [11].

References

1. Morgan, T., Rendell, L.E., Ehn, M., Hoppitt, W., Laland, K.N.: The evolutionary basis of human social learning. Proc. R. Soc. B 279653–66 (2012)
2. Premack, D., Woodruff, G.: Does the chimpanzee have a theory of mind? Behav. Brain Sci. **1**(04), 515–526 (1978)
3. Baron-Cohen, S., Leslie, A.M., Frith, U.: Does the autistic child have a theory of mind? Cognition **21**(1), 37–46 (1985)
4. Dennett, D.: The Intentional Stance Cambridge. MIT Press, MA (1987)
5. Frith, C., Wentzer, T.: Neural hermeneutics. In: Kaldis, B. (ed.) Encyclopedia of Philosophy and the Social Sciences, pp. 657–659. London SAGE publications Inc., Thousand Oaks (2013)
6. Frith, C.: Social cognition. Philos. Trans. R. Soc. Lond. B Biol. Sci. **363**(1499), 2033–2039 (2008)
7. Emery, N.: Gaze following and joint attention in rhesus monkeys (Macaca mulatta). J. Comp. Psychol. **111**, 286–293 (1997)
8. Bratman, M.: Shared cooperative activity. Philos. Rev. **101**(2), 327–341 (1992)
9. Monasterio Astobiza, A.: La comunicación bidireccional entre la cognición social y la cognición moral. Thémata: Revista de Filosofía **61**, 29–47 (2020)
10. Kobayashi, H., Koshima, S.: Unique morphology of the human eye. Nature **387**, 767–768 (1997)
11. Perrett, D., et al.: Visual cells in the temporal cortex sensitive to face view and gaze direction. Proc. R. Soc. Lond. B Biol. Sci. **223**(1232), 293–317 (1985)

Neuro-Rights for Human-Machine Interaction

Txetxu Ausín[1(✉)], Aníbal Monasterio Astobiza[2], and Ricardo Morte[2]

[1] Institute of Philosophy, Spanish National Research Council (CSIC), Madrid, Spain
txetxu.ausin@cchs.csic.es

[2] Faculty of Philosophy, University of Granada (URG), Granada, Spain
amastobiza@ugr.es, ricardo63@autistici.org

Abstract. The increase in our interaction with robots and other artefacts highlights the ethical risks that this convergence in multi-agent systems produces, especially with regard to the privacy and integrity of brain data. As a result, new human neuro-rights are being proposed that call for mental integrity being preserved from unwanted interventions and for protection against the coercive use of technologies that may lead to psycho-physical harm, discrimination, and injustice. It seems prudent to establish basic and universal safeguards, in Human Rights terms, that establish a framework for these technologies and that can be incorporated into international and national fundamental rights legislation to establish guarantees for them.

1 Human Beings as Socio-Technical Systems

HUMAN beings are increasingly configured as socio-technical systems where all our inter-relationships are technologically mediated. We interact physically, cognitively, and even emotionally with technology (Human- Machine Interaction HMI). Teams of humans work with "robots" in a kind of symbiosis between human and artificial intelligence: Human-in-the-loop. These convergent and disruptive technologies are also blurring the boundaries between human subjects and technology (Multi-Agent Systems). In this sense, values traditionally reserved for humans (such as autonomy, responsibility, creativity) are beginning to be projected onto technological entities, often driven by our tendency to anthropomorphism. [See INBOTS video on interactive robotics: *Robots are coming out of the factories. Are we ready?*: https://www.youtube.com/watch?v=h0QjNC8XQ1U].

2 Some Essential Risks

In relation to this human-machine interaction, it is possible to point out some questions and issues around autonomy, personal identity, psycho-social effects, dual-use of technology, security, responsibility, privacy, and non- consensual access to brain data [2].

Two main axes of risk are:

This paper benefits from Grant Agreements 780073-INBOTS and 779982- EXTEND. T. Ausín also belongs to BIODAT research project (Seneca Foudation) and *Detection and elimination of bias in triage and location algorithms for COVID-19* (BBVA Foundation).

M. A. Grau Ruiz (Ed.): INBOTS 2021, BIOSYSROB 30, pp. 25–29, 2022.
https://doi.org/10.1007/978-3-031-04305-5_5

(a) The use of Artificial Intelligence in the processing of brain data.
(b) The clear proximity of technology to our bodies (invasiveness)—the "Internet of Bodies", with an unprecedented number of connected devices and sensors attached to or even implanted and ingested into the human body. This has turned the human body into a technology platform. The IoB generates vast amounts of biometric and human behavioural data.

When an individual uses a technology such as brain- computer interfaces, the risk of *brainhacking* or *brainjacking* is possible. Both *brainhacking* and *brainjacking* involve a security breach in the software controlling the technology. *Brainhacking* involves disrupting the proper functioning of the algorithm behind the operations of a technology, such as brain-computer interfaces. *Brainjacking* involves the remote control of technology and can have repercussions on the control of a peripheral technological device such as a wheelchair, brain implant, digital medicine, etc.

We can imagine scenarios where damage is caused by the action of a technological device and this has not been controlled by the user of the brain-machine interface, by the user's brain in short, but by someone who has attacked and manipulated the algorithm that ultimately controlled the technological device. A malicious actor can take control of the technology, in this case a brain-computer interface and its prosthesis, and cause harm to the user or third parties. Ensuring the privacy and integrity of brain data is therefore crucial.

The other major risk axis of these technologies is invasiveness. They are in direct contact with our bodies, in particular with an organ essential to our personal identity: the brain and the nervous system in general. Many require a direct interface or very close contact with our brain, and this poses great risks. For example, the most reliable brain-computer interfaces are those that allow brain implantation, but this involves a surgical intervention that carries dangers of infection or toxicity to the nerve cells in the electrodes.

3 New Human Rights: Neuro-Rights

This new robot-human ecosystem has extensive applications and the potential to change our lives, rights, and fundamental freedoms. We have used the terms dangers and threats interchangeably to refer to these new risks linked to disruptive technologies that are beyond probabilistic determination and which, linked to human action, can acquire a catastrophic character, radically transforming the world and the very meaning of what is human.

Human Rights are traditionally considered to be subjective rights as normative positions or statuses that individuals possess in our relations with others. It seems obvious that an indispensable condition for developing our lives is moral agency and freedom of conscience, as stated in Article 18 of the 1948 Universal Declaration of Human Rights: "Everyone has the right to freedom of thought, conscience and religion...". Likewise, Article 12 of the same Declaration states that "No one shall be subjected to arbitrary interference with his privacy...".

However, the new technologies of human-machine interaction open the door to the dangers of manipulation and direct intervention on our brain and nervous system, threatening our mental integrity and, with it, the last bastion of our identity and our intimacy: our conscience.

These new threats call for an updating of Human Rights, understood in this way from an evolutionary perspective and adaptive to the new techno-scientific reality.

Ienca and Andorno [1] have therefore defended the definition of new Human Rights, Neuro-rights, in order to protect mental states from possible intrusions and manipulations:

- Right to cognitive freedom:

  This refers to "mental self-determination" and would be, in a way, an extension of the above-mentioned human right to freedom of thought and conscience. It includes two closely related concepts: the right of individuals to use emerging technologies; and the protection of individuals against the coercive use of these technologies and the possibility that the technology may be used without their consent.
- Right to mental privacy:

  If we consider the problems in achieving adequate protection of the right to privacy, it seems clear that standards need to be adapted to achieve the same kind of protection for mental privacy. It means protecting individuals against the illegitimate use of their brain information and preventing possible leaks of this data.
- Right to mental integrity:

  Intrusions or actions on a person's brain can create not only a violation of their privacy but also a detrimental change to their neural status. The presence of harm is a necessary condition for a violation of the person's mental integrity to have taken place.

  In a similar vein, the group led by neuroscientist Rafael Yuste, director of the BRAIN project, has formulated a proposal for human neuro-rights (https://nri.ntc.columbia.edu/content/spanish-webpage):
- Right to personal identity:

  Limits must be developed to prohibit technology from altering the concept of self. By connecting individuals' brains to computers, technology could blur the line between a person's consciousness and external inputs.
- Right to free will:

  People should be able to make and have control over their own decisions, without manipulation by unknown external technologies.
- Right to mental privacy:

  All data obtained after measuring neural activity (NeuroData) must be kept private. In addition, the sale, commercial transfer and use of neural data must be strictly regulated.
- Right to equal access to brain enhancement:

  Guidelines regulating the development and applications of brain enhancement technologies should be established at international and national level. These guidelines should be based on the principle of justice and ensure equal access for all citizens.
- Right to protection against bias:

Countermeasures to combat bias should be established as standard in machine learning.

Both proposals agree on the basics and can be summarised in two mandates: Mental integrity must be preserved from unwanted interventions, as the ultimate bastion of personal identity and freedom. And this freedom must be protected against coercive use of technologies that can lead to psychophysical harm, discrimination, and injustice.

It is a step in the right direction that the Republic of Chile has passed a law on neuro-rights. Chile became the first country in the world to protect neurorights. On 29 September 2021 The Chamber of the Deputies endorsed the report of the joint committee on the bill enshrining physical and mental integrity and neural inviolability. An amendment to the constitution was created for a law to protect mental integrity from the advance of neurotechnologies.

The Chilean constitutional reform project is based on the need to establish a new fundamental right to protect the physical and mental integrity of individuals from technological mechanisms that seek to "increase, diminish or disturb individual integrity without due consent".

Chilean legislators aim to protect neurorights and mental integrity in the face of present and future advances in neurotechnologies (Bulletin 13,827–19). The Chamber of Deputies said in a statement that "Chilean law establishes that scientific and technological development must be at the service of people and that it must be carried out with respect for life and physical and mental integrity".

Spain, on the other hand, in its new Digital Rights Charter introduces a section on neurotechnologies with explicit mention of neurorights, and while it's a non-binding framework, it may inspire new legislation.

Neurorights are not born in a vacuum. Throughout the history of philosophy and legal thought, various concepts and constructs have been developed that have served as the basis for neurorights. Although it is true that neurorights have moved very quickly from being analysed in the ethical debate, advocacy and from there to politics; there are still unanswered questions and an open future.

How neurorights are to be named, defined, interpreted and enforced remains to be decided. Different commentators and authors argue that it remains to be decided how to interpret the neurorights. Some consider that a decision will have to be taken on whether neurorights will be seen as international human rights law, should be interpreted as entirely new human rights or as evolving interpretations of existing rights. However, it is not uncommon to observe an evolution in regulation in the face of advancing technology and science. Well known example is the change in the concept of brain death with the introduction of the mechanical ventilator.

In any case, we believe that neurorights are the best tool to guarantee the protection of people in the face of the progressive advance of neurotechnologies that can intervene and modify the human brain.

Let us hope that neurorights will expand or be accommodated within the existing human rights framework and guarantee the protection of human dignity in the face of advancing AI, genomics and neuroscience.

4 Conclusion

For the time being, the proposal of these new human neuro- rights opens up an absolutely necessary debate on the growing role of technologies in our society and their enormous disruptive potential, to the point of being able to transform our very sense of self. It seems prudent to establish basic and universal safeguards, in terms of Human Rights, that establish an insurmountable framework for these technologies and that can be incorporated into international and national fundamental rights legislation in order to establish guarantees for them.

It seems clear that the right way to design and implement the safeguards mentioned above is to apply the criteria of Privacy by Design and by Default, and the same applies to Ethics. In order to advance along this path, an appropriate instrument would be the use of "impact assessments" in the early stages of the design of any technology. It is worth mentioning the Data Protection Impact Assessment introduced in Article 35 of the General Data Protection Regulation (GDPR: https://gdpr.algolia.com/es/gdpr-article-35) or the recently published proposal for an "ethical impact assessment" model in the field of health [3] which proposes the following principles: explainability, autonomy, beneficence, non-maleficence, fairness, sustainability, precaution, privacy, democracy, controllability and auditability.

References

1. Ienca, M., Andorno, R.: Towards new human rights in the age of neuroscience and neurotechnology. Life Sci. Soc. Policy **13**, 5 (2017)
2. Monasterio, A.A., et al.: Traducir el pensamiento en acción: Interfaces cerebro-máquina y el problema ético de la agencia. Revista de Ética y Bioderecho **46**, 29–46 (2019)
3. Morte Ferrer, R.: Reflexiones sobre las evaluaciones de impacto. Una propuesta para un modelo de Evaluación del Impacto Ético en el ámbito de la salud. Dilemata **32**, 71–82 (2020)

Rethinking Environmental Vulnerability

M. Funk(✉)

Chair of Philosophy of Media and Technology, Institute of Philosophy, and the Research Group Cooperative Systems, Faculty of Computer Sciences, University of Vienna, Vienna, Austria
michael.funk@univie.ac.at

Abstract. Human-technology relations gain a specific interest in the philosophy of technology. Recently these relations have been challenged by ecological problems – concerning climate change and natural environments – as well as privacy – concerning the fair use of personal data and information protection. Both aspects stand for existential relations that bear certain vulnerabilities. In this paper I will analyze from a philosophy of technology point of view existential human–environment-human relations with a specific focus on the robotic Anthropocene. On a material bodily level vulnerability shapes the ways we use robots in natural and cultural environments (ecology); which is recently supplemented by voice operating technologies on a non-material level. Especially the later ones directly affect privacy and the existential vulnerability of human dignity when it comes to linguistic interactions with robotic systems. I argue that vulnerability is a fundamental aspect of Robot Ethics. It is not to be confused with a mere handicap since it enables human creativity and flexibility – in contrast to machines.

1 Introduction

I want to make the point, that human existence is not only incidentally based on vulnerability. Instead, vulnerability is a basic condition of both human creativity and the capabilities of flexible adaptation to new environments.

In the age of the Anthropocene [1], vulnerability receives an increased self-referential status due to its environmental feedback relations. On the one hand this self-referentiality is as old as human material culture, since humans for millions of years have been treating their environments with more or less complex tools like hand axes [2]. Technically shaped environments turn into cultural environments with a certain effect on other humans – for instance new generations that learn techniques of how to interact within culturally shaped situations in their childhood. Environments shape us as we shape environments [3]. On the other hand, the quality of human–environment-human interactions seems to be heavily unbalanced in what we call Anthropocene. Human activities became the primary factor of global developments in the recent epoch of earth history – including Geoengineering, Carbon Dioxide Emission, global warming etc.

This work has been supported by the European Union's Horizon 2020 research and innovation program under the CSA project "INBOTS – Inclusive Robotics for a better Society" (grant agreement No 780073).

M. A. Grau Ruiz (Ed.): INBOTS 2021, BIOSYSROB 30, pp. 30–35, 2022.
https://doi.org/10.1007/978-3-031-04305-5_6

Beside others, robotics became one of the most significant emerging technosciences [4] within the broader developments of the Anthropocene. A characteristic trademark of the peculiar robotic Anthropocene is the new interface of voice operating technologies. Our environment is getting into dialogues with us [5]. This aspect bears a certain relevance because the vulnerability [6] of material bodies in human–environment-human interactions is supplemented by non-material vulnerability of voice acts. The following analysis consists of a propaedeutic part "Terms and Methods" and a second one about "Argument and Discussion".

2 Terms and Methods

2.1 Environment

The term "environment" involves two main dimensions: a) natural environment and b) cultural environment. When it comes to the ecological perspective with respect to biological species, we are used to talk about natural environments (= ecology in the narrow sense). Products of human actions, in contrast, classically belong to culture and therefore shape a very specific form of cultural environments ($\neq$ ecology in the narrow sense). However, I want to suggest a broader understanding of ecology that involves both natural and cultural aspects of human life [7]. Therefore I follow certain body-oriented approaches that critically deconstructed the classical Aristotelian divide of nature and culture [8]. It is due to human nature, that bodily dispositions are turned into cultural practices within natural-cultural environments (= ecology in the broad sense). Human bodies are then not the opposite of environments but already embedded into ecological situations [9]. Mediated through embodied actions environments shape us as we shape environments.

2.2 Vulnerability

Vulnerability is an existential characteristic of human life, meaning the fragility of physiological and mental health (body), of communication and social life (misunderstandings, communication breakdown, clash of cultures) as well as the tentative and incomplete status of human knowledge. Usually vulnerability is linked to finitude – the unavoidable ending of human lifetime [10]. Finitude has been questioned in mythological tales for thousands of years. Today it is the object of trans- and posthumanist speculations [11]. However, finitude is not the topic of this paper.

2.3 Methods

The analysis of vulnerability is based on (post)phenomenological methods [12] including a relational approach of human-technology-worlds relations [13]. Embedded into a materialistic basis, expanded hermeneutics [14] then becomes the method of bodily interpretations including constructivist elements of technoscientific constructions of mediated, instrumental realities [15]. Vulnerability is therefore conceptualized on the basis of ordinary life actions. Also its existential dimension is understood as a relational – not relativistic (!) – culturally embedded practice [16]. In particular Ihde's approach is to be supplemented by a Wittgensteinian perspective on linguistic techniques and its implemented communicative vulnerabilities [5, 17, 18].

3 Argument and Discussion

The general argument considers human–environment-human relations in a way that humans are shaped by their (cultural-social, linguistic and material + natural) environments as they shape their environments in return. There is so to say a triangle of individual-society-nature. On the one hand human life can be understood as risky, endangered and fragile. However, it is not sufficient to conclude with a "Mängelwesen" argument as it has been developed by Herder and later on by Gehlen. Vulnerability goes hand in hand with flexibility and potentiality of adaptation. If humans would not be vulnerable they would be able to creatively challenge problems in real life situations. In a nutshell the message for treating challenges of ecological developments like climate change is the necessity to accept vulnerability and to prevent harming vulnerabilities in an existential way. Therefore the slogan "Inclusive Robotics for a Better Society" can be turned into "Better Robotics for Vulnerable Societies". We need to conserve the conditions of human life in that way. Of course there is no doubt that progress in medicine, not only in pandemic situations, is of priority and an outstanding value. Nevertheless, medical technologies like care-robots should be understood as a contribution to the extension of healthy and happy – although still vulnerable – lifetime, which is not to be confused with the trans-/posthumanist dream of overcoming finitude and vulnerability [19].

Any attempt to create infinite super-intelligence or so called "inclusive robots" as a technocratic benchmark of humanity will run into a contradiction. The point is that the overlooking of vulnerability and finitude becomes an overlooking of real problems that have to be challenged by free and creative humans. In that sense not a vision of "inclusive robots" but of inclusive human–environment-human relations seems to be an appropriate policy for nature conservation – especially when it comes to long term-responsibility and appropriate conditions of a happy and dignified life of future generations. Therefore it is required to accept and rationally prevent conditions of vulnerable existence. Once more: Of course progress in medicine is needed in order to prevent as much physiological or psychological harm as possible. The argument here is more about our sociocultural image of robotics. These machines are not the Holy Grail of inclusion. In the next step, we should raise the question of inclusive human–environment-human relations independent of robotics and just in the over next step it is to asked, what robots can contribute as mere means to fulfill ends in the context of vulnerable human life.

In that sense, that is the crucial point, privacy appears to be a problematic issue in many ways. Why do we need privacy? Because it is both a condition and a consequence of vulnerability. Privacy should not be reduced to a mere side-effect of current (post)digitization. Instead it is a deeply entangled aspect of human existence [20], so to say the solipsistic self-relation everybody has within her/his concrete human–environment-human relations. If this relational aspect is lost, we lose our power to face the many challenges we are confronted with – starting with the huge ecological challenges of our days but also the resilience of democratic societies. In a nutshell, Carissa Véliz argues:

"It's too late to prevent the data economy from developing in the first place – but it's not too late to reclaim our privacy. Our civil l liberties are at stake. The decisions we make about privacy today and in the coming years will shape the future of humanity for decades to come." [20, pp. 3–4].

The question concerning privacy is indeed a bodily one. For instance smart watches and the measurement of physical issues like the heart rate, temperature etc. can be linked to some kind of "emotional surveillance" – since they are used as indicator for a certain emotional state. Both bodily-material and bodily-emotional privacy are entangled due to to new technological means. In this context Véliz concludes:

"Surveillance is not only about what you *do*, but about what you *think* and *feel* – it's under-the-skin-surveillance. Your body is scrutinized to infer both your emotions and the state of your health." [20, p. 203].

Human–environment-human relations are bodily relations. In a peculiar way, its specific vulnerability is today shaped by surveillance technologies that call for sustainable treatment of personal data in fate of privacy. For Véliz the message is clear: "Refuse the unacceptable. Take back control of your personal data, and privacy will prevail." [20, p. 208] Thereby the author stands in line with others like Shoshana Zuboff who has been critically analyzing structures of "surveillance capitalism" [21]. In political terms human–environment-human relations receive another ethical brisance. It's not only about control of personal data, it's also about the societal environments we want to create with and without AI, robots – or any other computer- and algorithm-based technology. In her influential book *Weapons of Math Destruction* Cathy O'Neil argues:

"Data is not going away. Nor are computers–much less mathematics. Predictive models are, increasingly, the tools we will be relying on to run our institutions, deploy our resources, and manage our lives. But [...] these models are constructed not just from data but from the choices we make about which data to pay attention to–and which to leave out. Those choices are not just about logics, profits, and efficiency. They are fundamentally moral." [22, p. 218].

Our moral choices are fragile and vulnerable itself. They might fail. However, that is what makes us become humans: We have the choice to perform inclusive societies – with or without (inclusive)robotics. This kind of freedom is both a matter of creativity and responsibility. On a bodily level environmental vulnerability calls for its translation into actions that transform relations in a morally and ethically desirable way – so to say the material "embodiment". Therefore, Mark Coeckelbergh argues:

"Like all performance, political performance is never fully scripted and determined; it is also a matter of rehearsing and trying things out. And neither the success nor the performative *power* of a performance is ever guaranteed. As vulnerable human beings in complex social-natural environments, we have to perform in the face of risk and uncertainty." [23, p. 156].

4 Conclusion

We do not not need vulnerability. We are vulnerable. What we need is a sharpened awareness of better robotics as mere means within inclusive fragility. It is about our human–environment-human relations that come first and set the stage for privacy and ecological responsibility. Robotics comes later. Any attempt to replace risk and uncertainty by predictive mathematics/algorithms would lead to the end of political and social life. Do we want that?

References

1. Steffen, W., Crutzen, P., McNeill, J.: The anthropocene. are humans now overwhelming the great forces of nature. J. Human Environ. **36**(8), 614–621 (2007)
2. Schrenk, F.: Paläoanthropologie. In: Bohlken, E., Thies, C. (eds.) Handbuch Anthropologie. Der Mensch zwischen Natur, Kultur und Technik, pp. 197–207. J.B. Metzler, Stuttgart (2009)
3. Funk, M., Frauenberger, C., Reichl, P.: Vom künstlichen Leben zur Lebenskunst – Was die Ethik digitaler Bildung von ökologischer Verantwortung lernen kann. In: Medienimpulse, Bd. 58, Nr. 03 (2020). https://doi.org/10.21243/mi-03-20-17
4. Jordan, J.: Roboter. Berlin University Press, Wiesbaden (2017)
5. Coeckelbergh, M.: Using Words and Things. Language and Philosophy of Technology. Routledge, New York & London (2017)
6. Coeckelbergh, M.: Human Being @ Risk. Enhancement, Technology, and the Evaluation of Vulnerability Transformations. Springer, Dordrecht (2013)
7. Janich, P.: Sprache und Methode. Eine Einführung in philosophische Reflexion. Tübingen, Francke (2014)
8. Ihde, D.: Bodies in Technology. University of Minnesota Press, Minneapolis (2002)
9. Fuchs, T.: Verteidigung des Menschen. Grundfragen einer verkörperten Anthropologie. Frankfurt a.M., Suhrkamp (2020)
10. Rentsch, T.: Heidegger und Wittgenstein. Existenzial- und Sprachanalysen zu den Grundlagen philosophischer Anthropologie. Klett-Cotta, Stuttgart (2003)
11. Came, D.: Der Tod und seine Leugnung im Transhumanismus. In: Göcke, B.P., Meier-Hamidi, F. (eds.) Designobjekt Mensch. Die Agenda des Transhumanismus auf dem Prüfstand, pp. 95–113. Herder, Basel (2018)
12. Ihde, D.: Postphenomenology. Essays in the Postmodern Context. Northwestern University Press, Evanston (1993)
13. Ihde, D.: Technology and the Lifeworld. From Garden to Earth. Indiana University Press, Bloomington (1990)
14. Ihde, D.: Expanding Hermeneutics. Visualism in Science. Northwestern University Press, Evanston (1998)
15. Ihde, D.: Instrumental Realism. The Interface between Philosophy of Science and Philosophy of Technology. Indiana University Press, Bloomington (1991)
16. Coeckelbergh, M.: Growing Moral Relations. Critique of Moral Status Ascription. Palgrave MacMillan, Basingstoke (2012)
17. Coeckelbergh, M., Funk, M.: Wittgenstein as a philosopher of technology: tool use, forms of life, technique, and a transcendental argument. Human Stud. **41**(2), 165–191 (2018)
18. Funk, M.: Repeatability and methodical actions in uncertain situations: wittgenstein's philosophy of technology and language. In: Coeckelbergh, M., Funk, M., Koller, S. (eds.) Techné: Research in Philosophy and Technology. Special Issue, vol. 22, no. 3, pp. 351–376 (2018). https://doi.org/10.5840/techne201812388
19. Lorenz Sorgner, S.: Was wollen Transhumanisten? In: Göcke, B.P., Meier-Hamidi, F. (eds.) Designobjekt Mensch. Die Agenda des Transhumanismus auf dem Prüfstand, pp. 153–179. Herder, Basel (2018)
20. Véliz, C.: Privacy is Power. Why and How You Should Take Back Control of Your Data. Bentam Press, London (2020)

21. Zuboff, S.: Das Zeitalter des Überwachungskapitalismus. Frankfurt a.M., New York Campus (2018)
22. O'Neil, C.: Weapons of Math Destruction. How Big Data Increases Inequality and Threatens Democracy. Crown, New York (2016)
23. Coeckelbergh, M.: Moved by Machines. Performance Metaphors and Philosophy of Technology. Routledge, New York (2019)

Work

The Ethical Debate on Human Work: A Labour Law Approach

Maria Luz Vega(✉)

Research Department International Labour Organisation, Geneva, Switzerland
vega@ilo.org

Abstract. The incorporation of machines into the world of work raises the need for their eventual labour law regulation. However, labour law is focused in human beings and must remain so in order to guarantee its purpose and essence. In this new vision, the protective essence of social regulation should not be lost at the risk of falling into an exacerbated individualism that would increase inequalities and could affect social justice.

1 Introduction: Aspects and Characteristics of Human Work

The idea of work as a human activity is developing and moving forward in various cultural and social areas as a result of theological, philosophical and political reflection. That in turn is being consolidated with the recognition and declaration of the existence of human rights, and the affirmation of freedom, equality and fraternity of men and women regardless of their race or social status.

Adam Smith or David Ricardo refer to "human" work, describing it as the only source of all wealth[1], while Hegel, some years later, underscores the idea of its human and transforming nature, considering that the human being exists genuinely through work, since it satisfies the human being's own needs and with those, the needs of others. In addition, Hegel reflects on the introduction of the machine in that context. In his view, that introduction not only led to the deterioration of working conditions, condemning some to poverty, but also limited the expression and exercise of human faculties, preventing integration into civil society. Work from that perspective gives meaning to life – it is not simply a means, but an end in itself.

Later, Karl Marx in Capital deepens the idea of work as a transforming action, defining it as an act that occurs "between man and nature". In the same work he indicates that although animals produce, they do so unilaterally to meet their immediate needs or the needs of their offspring, while a human does so from a place of freedom to transform nature, and not always or only to meet immediate needs. That is why, he concludes, work is constitutive of the essence of humankind.

Work is therefore a multidimensional activity that not only implies an effort but expresses an accumulated knowledge that involves the application of ideas and knowledge. It is that "subjective" implication that gives social recognition to men and women

[1] See Meda D. Le travail, un valeur en voie de disparition. Ed Aubier, paris (1995).

M. A. Grau Ruiz (Ed.): INBOTS 2021, BIOSYSROB 30, pp. 39–45, 2022.
https://doi.org/10.1007/978-3-031-04305-5_7

and causes them to differentiate themselves from the machine. Thus, the work done is exceedingly social (and therefore intrinsically human), insofar as it represents the individual and is an expression of ethical and cultural values.

For Supiot[2], work is distinguished from activities that correspond to an obligation, whether entered into voluntarily or legally. It is enough that a commitment to action be given legal effect (as it is the law that consecrates social utility) for that action to be described potentially as "labour". Therefore work is not only paid activity, but also volunteering or work that is traditionally unpaid but that has the impact and characteristics described (this includes, for example, work ranging from that of a housekeeper to that of interns).

Work is, thus, inseparable from human life (for that reason, it is difficult to establish apocalyptic theories about its end), and it is necessary, not only because it serves the production of goods or services, but because of its transforming capacity. In addition, it generates relationships and allows the person to "transcend", insofar as their production (output) will last through time. Therefore, as a social fact, it needs ethics and rules. Hence its regulation is necessary.

Work is (as already mentioned) a right, and more so a fundamental human right, as well as a duty that goes beyond obligation insofar as it is necessary for personal development. As Delos points out[3], "man enters the State as a subject of rights and duties. It is a natural quality and, far from being lost in the city, on the contrary, it is there where the quality manifests itself in all its value." The human being knows that they need to work, and that gives work a legal identity in the social sphere.

From very early on, the right to work was included in numerous Declarations and International Charters, even before being practically recognized in all national constitutions, since in general, labour law is included in charters as a consequence of the need for reconstruction after war (as is the case of France (1946), Switzerland (1947), Italy, Luxembourg (1948), etc.)[4]. Let us not forget that unlike the constitutions sanctioned in the nineteenth century, (which following the liberal model only considered the rights of the individual without taking into account their position in society), the social constitutions of the twentieth century, by adding new rights, integrate the position of the individual in society, essentially on the basis of their character as a worker, thus placing work at the centre of politics and the state.

Article 23 of the Universal Declaration of Human Rights[5], considered the foundation of international standards on the subject, states that "everyone has the right to work, to

---

2 Supiot A. (Under the direction of) Au délà de l'emploi. Paris Flammarion. 1998.

3 Delos JT et al., Los fines del derecho: el bien común, seguridad jurídica, en los fines del derecho, México UNAM, 1975, page 51.

4 Incorporation into the constitutions is the result of so-called social constitutionalism, a universal movement that defends and promotes the incorporation of social rights at the highest level and that began with the adoption of the Constitution of Mexico of 1917 (a direct result of the Mexican Revolution); in Argentina, with the Constitution of the Province of Mendoza in 1916; in Germany, with the Constitution of the Weimar Republic in 1919; and in Spain with the Spanish Constitution of 1931.

5 http://www.un.org/es/universal-declaration-human-rights/.

free choice of employment, to just and favourable conditions of work and to protection against unemployment."

For its part, Article 6 of the International Covenant on Economic, Social and Cultural Rights[6], states that "the States Parties to the present Covenant recognize the right to work, which includes the right of everyone to the opportunity to gain his living by work which he freely chooses or accepts, and will take appropriate steps to safeguard this right", reiterating the idea of self-determination and humanity linked to work.

Completing this panorama, numerous regional declarations of human rights emphasize the same idea, reiterating the right/duty dichotomy and deeming it fundamental and human.

From the perspective set out it seems impossible that machines, even the most sophisticated artificial intelligence (AI) and/or robot systems, could enjoy a right that is human by nature – although the subject seems to be subject to debate. For example, in October 2016, Saudi Arabia granted Saudi nationality to Sophia, the humanoid developed by Hanson Robotics[7], directly ascribing to it the status of humanity.

In contrast, a group of experts in robotics (200 experts from 14 countries), faced with the European Union (EU) proposal to draft a directive to give legal status to robots (see below), recently sent an open letter to the EU[8], warning that such a project could go against Human Rights and the international instruments that sustain them. The group of experts expresses its rejection of the directive on Robotics and Artificial Intelligence (preparation under way) granting "personhood" to machines, as advocated in the resolution on the rules of civil law of robotics approved by the European Parliament in 2017 (and that is serving as a basis for the new regulation). The signers warn that granting the status of "electronic people"[9] to robots would mean recognizing that they have rights and responsibilities, with the economic, legal, social and ethical impact that implies.

The question is therefore to determine, if the right to work is exclusively human, what regulation is necessary for the activity of robots? Do machines have rights with respect to the work they do? Is it necessary to reform labour law? Those are some of the reflections that we intend to address in the debates on the future of work.

2 AI, Robots and Machines: Do They Have Rights?

The fundamental premise in the debate is to consider which are the ethical criteria applicable to beings or entities other than humans. Although the relationship between ethics and law have been discussed by the great thinkers (for example Kant or Kelsen), they always appeared to refer to the human being, considering that ethics is part of

---

[6] https://www.ohchr.org/SP/Professionalinterest/Pages/cEScr.aspx.

[7] Drawing numerous remarks in relation to a country where the rights of part of the population (women) are limited.

[8] http://www.robotics-openletter.eu/.

[9] The European Parliament introduced the idea of creating a status or legal personality for machines (paragraph 59(f) of its resolution on the civil law rules of) with the idea that "at least the most sophisticated autonomous robots can establish themselves as electronic people responsible for repairing any damage they may cause.".

the social relationship and coexistence and is only fitting for the Aristotelian "Zoon politikon".

However, some more recent trends extend the idea of ethics to other beings[10] because the purpose is to achieve a society with common values that encompass all subjects and that is therefore more inclusive. The same author sketches the idea that intelligent machines could/should have moral standards. The discussion lies again in the hypothesis (already mentioned) on the need to have consciousness and free will. It is clear that the machine will never be a "moral agent" (someone who acts consciously), although today it could be understood, by the dynamics of the facts, that there is another type of ethics built by the social dynamics that govern the interactions and relationships between various entities and beings. That is, as Mark Coeckelbergh indicates[11], that a moral standard does not refer to the entity, but to "us" and to the existing relationship between us and the entity. However, that theory has not been widely accepted and has various aspects that can be discussed.

In fact, machines having rights is not simply a matter of applying moral considerations to non-human entities, but of examining the consequences of their interaction with humans since this interaction can alter society and therefore ethics in itself.

The EU is working to that end. Indeed, the introduction of the aforementioned proposal for recommendations to the Commission on civil law rules on robotics[12] indicates the importance of "the legislator weighing the legal and ethical consequences without hindering innovation", even more so considering that "in view of the increasing social division and the decline of the middle class, it should be borne in mind that the progress of robotics could translate into a high concentration of wealth and power in the hands of a minority"[13].

For the European Parliament "any evolution of the machine must preserve the dignity, autonomy and self-determination of the individual"; it therefore considers "The Union could play an essential role in establishing basic ethical principles to be respected in the development, programming and use of robots and AI and in the incorporation of such principles into Union regulations and codes of conduct, with the aim of shaping the technological revolution so that it serves humanity and so that the benefits of advanced robotics and AI are broadly shared, while as far as possible avoiding potential pitfalls"[14].

There is no doubt about the need to establish certain ethical rules for machines, which were already outlined somewhat by Asimov in his laws of robotics (1942)[15]. Moreover, the European Union considers it necessary to legislate and go beyond ethics, considering that today's machine "presupposes a certain intelligence". In fact, the first premise of the eventual regulation is the need to find a definition of intelligent robot, because only

---

[10] Anderson, S. (2008) Asimov's "Three Laws of Robotics" and machine metaethics (4) 477–493, page 480.

[11] Coeckelberg, M. (2012) Growing Moral Relations: Critique of Moral Status Ascription New York Palgrave Macmillan, page 25.

[12] European Parliament resolution of 16 February (2015/2103 (INL)) http://www.europarl.europa.eu/sides/getDoc.do?pubRef=-//EP//TEXT+TA+P8-TA-2017-0051+0+DOC+XML+V0//ES.

[13] Paragraphs B and K of the introduction.

[14] Op cit.nota 18 paragraphs O and V.

[15] Asimov I. Handbook of Robotics, 56th Edition, 2058 A.D.

with an identification of the subject and object of the norm can the rules of its behaviour be regulated and, therefore, the ethical principles that must govern it. For that reason, it proposes a legal personality of robots (see above letter from experts on the electronic personality) that allows the robots to be responsible for the damage they may cause and that covers the assumptions in which the robots make intelligent autonomous decisions or interact with third parties independently. At the same time, it makes them subjects of taxation in order to avoid that aforementioned concentration of wealth in the hands of a minority.

All the issues mentioned by the EU are debatable, as the experts themselves indicate in the open letter (see above), but the truth is that regulation and more sophisticated norms to regulate the behaviour of machines appear necessary.

Objectively, the robot does not work in its own right; it is useful in the facilitation or the carrying out of work. That notwithstanding, the possible level of autonomy in the most sophisticated entities makes it necessary to establish limits, criteria and rules.

Is express regulation necessary? What type of regulation would be necessary? When should it be regulated?

The central problem is defining the relationship between the human being and the machine and the real reason, if there is one, for the independence of the mechanical entity. Some authors emphasize the idea that we cannot talk about AI[16] or robotization without placing the human being at the centre, and that this implies a change of perspective on work (which captures Plato's "Protagoras": man "as the measure of all things"). The goal is no longer efficiency, but creativity and the power to exercise a new way of life that is more based on the development of "being" and not on "producing".

Productive and efficient work, subject to current labour legislation, would be transformed in the future due to interactions with "machines". The goal of work would not so much be to produce; instead it would be the way in which the human being is realized[17]. The idea would be to give meaning to the activity, focusing on the representation of happiness and freedom, as well as commitment to the work done. The machine would be a contributor to the work to be carried out and a basic instrument that would make it possible to adjust the life/work cycle, allowing more space for free time.

Human beings must remain the object of labour regulation, but the existence of a developed AI will modify the regulation and the meaning of said regulation. In this new vision, the protective essence of social regulation should not be lost at the risk of falling into an exacerbated individualism that would increase inequalities and could affect social justice.

For some that perspective may be of an overly visionary techno-optimistic nature, but the truth is that the Ius laboralista doctrine mainly considers that productive work as the object of regulation is on the way to extinction, but does not define what kind of work would take its place.

[16] Artificial intelligence describes the work processed by machines that would require intelligence if it were performed by humans.

[17] Scrivens, Jeremy. Putting people first in the future of work, liberates the flow of joy https://www.Linkedin.com/pulse/.

In that regard, it is not possible to speak of, to paraphrase Rifking[18], the end of work, but instead of a deep transformation far removed from the productive work (and to some extent manual work) that gave rise to the class struggle. The concept today includes a new form of productive expression of a more general and integrative nature since it is presumed that the work of the future will be more ethical and more equal with the benefit of the use of technology. Human beings should not face off with machines, but make it part of their labour development, avoiding the risk of absolute dependence on technology that could deprive them, to a certain extent, of their freedom.

3 Human Work and Social Justice: The Foundation on Which to Build the Future

The implications at all levels of possible policy decisions on the different subjects raised aboved are much more extensive than the current debates on labour law reform. There is no doubt that it transcends the very concept of the welfare state and the social and relational model. The labour relationship that has to date been defined as the basis of the law has culminated in the current state of inadequacy. Labour regulations should include many relationships and activities, including the aspects related to the use of machines.

The human being in the twenty-first century faces different challenges and has different ways and means of resolution. The idea of a new social contract hangs in the air and in discussions, as a cross-cutting foundation for the actions to be undertaken. In that context, however, it is clear that catastrophic and apocalyptic predictions are pointless. However, the need for (at least a minimum of) protection remains a basic requirement and appears to go beyond the traditional concept of worker as there are new elements that are included in this equation.

The world is changing but social justice remains a goal. Men and women seek peace, well-being and a modus vivendi that allows them to feel fulfilled. That is the challenge that we face today and the great question surrounding the future of work. Only human beings can decide their future, using the tools at their disposal and maintaining the priorities that guided peace and progress over the past century. Technology must always be used in a targeted way so as to be advantageous, and any new contract, guidance or policy should safeguard that way forward and build on already established values.

The fundamental question is to determine the role of work in the economy and in society, as well as its value in the development of human relations. The aforementioned new social contract can only be based on already established gains.

The world may change but human beings needs peace, justice and equity. Those are the basic values that should guide the new regulation and that are present in the basic texts that inspire social regulation. A new labour law and new ways to achieve its goals may be needed; however, the new social contract must preserve human beings from unnecessary risks and give them the ability to express their needs and potential with basic safeguards for their character and dignity.

---

18 Rifkin. *The End of Work: The Decline of the Global Labor Force and the Dawn of the Post-Market Era*. USAG. P. Putnam's Sons. 1995.

References

1. Anderson, S.: Asimov's "three laws of robotics" and machine metaethics. (4), 477–493 (2008)
2. Asimov, I.: Handbook of Robotics, 56th edn. 2058 A.D
3. Coeckelberg, M.: Growing Moral Relations: Critique of Moral Status Ascription. Palgrave Macmillan, New York (2012)
4. Delos, J.T., et al.: Los fines del derecho: el bien común, seguridad jurídica, en los fines del derecho, México UNAM (1975)
5. Meda, D.: Le travail, un valeur en voie de disparition. Ed Aubier, Paris (1995)
6. Rifkin. The End of Work: The Decline of the Global Labor Force and the Dawn of the Post-Market Era. USAG. P. Putnam's Sons (1995)
7. Scrivens, J.: Putting people first in the future of work, liberates the flow of joy. https://www.Linkedin.com/pulse/
8. Supiot, A.: (Under the direction of) Au délà de l'emploi. Paris Flammarion (1998)

Check for updates

Digitalisation: Shaping the Way We Work

Roberto Suarez-Santos and Luis Rodrigo Morales-Velez(✉)

International Organisation of Employers, Geneva, Switzerland
Morales@ioe-emp.com

Abstract. Technological challenges have dramatically impacted the world of work. Acceleration of the impact of technology through digitalization, robotics, and artificial intelligence in the way we work are putting pressure our policy agendas, and both governments and businesses stand to benefit from this technological revolution. The adoption of enabling policies will outline the pace of adoption of these technologies towards development and innovation. Robots can make our lives better and more efficient and connectivity has brought enormous opportunities which are being multiplied by emerging technology breakthroughs. Taking full profit of the digital era will require enabling environments for digital expansion and inclusiveness through appropriate policies. The digital divide should be addressed by governments and Technological skills would need to be attained fast. Meeting the needs of industry and the rapid changes in the world of work, demands that education systems be reformed. We have to do better on filling the skills gaps to develop employability in a world where technology and robotics so demand.

1 Introduction

Technological advances have dramatically impacted the world of work. No one disputes this point, but without a doubt, we are, and we will continue to witness the massive impact of technology, be it digitalisation, robotics, or artificial intelligence.

The Covid-19 pandemic has accelerated the impact of technology. David Autor -a leading economics professor at Massachusetts Institute of Technology (MIT) and co-chair of the MIT Work of the Future Task Force- has called the Covid-19 crisis an "automation forcing event" that will fundamentally transform the economy.

Technology and innovation have contributed to business continuity and resilience during the pandemic. Imagine this pandemic without connectivity, remote work, or e-commerce, and without all distancing measures that were only possible thanks to digital means. They have helped to save jobs and maintain productivity and prevent contagion.

As social distancing measures became mandated and rules to reduce physical presence at work prevailed, many businesses turned to digital technology to continue operating. They either expanded pre-existing telework capabilities or rapidly adopted and implemented policies to maintain production levels while working remotely. Some 20% of the workforce, on average, could work remotely. This percentage is four to five times higher than before the pandemic [1]. E-commerce transactions also grew two to five times in many economies compared with pre-covid. Other kinds of virtual transactions such as

M. A. Grau Ruiz (Ed.): INBOTS 2021, BIOSYSROB 30, pp. 46–50, 2022.
https://doi.org/10.1007/978-3-031-04305-5_8

telemedicine, online banking, streaming entertainment - to give a few examples, took off. Online doctor consultations through web-based applications grew exponentially, with an expected increase of 72% by 2024 [2]. In China, e-commerce, delivery, and social media jobs grew by more than 5.1 million during the first half of 2020. This rise helped save lives, save livelihoods, and even limit the mental health effects of lockdowns.

COVID-19 may have accelerated the adoption of automation and artificial intelligence, especially in jobs with high physical proximity. In a recent survey, two-thirds of 800 senior executives said they were stepping up investment in automation and artificial intelligence [3]. By June 2020, production figures for robotics in China exceeded the levels before the pandemic. Indeed, we cannot ignore the positive impact technology has had on our societies. Still, we must recognise the constant apprehension that the new technologies will disrupt labour markets, increase inequalities, and reduce well-being. They should not hold back innovation and action yet require examination and analysis. As in previous revolutions, the Fourth Industrial Revolution can raise global income levels and improve people's quality of life across the globe.

2 Shaping the Way We Work

Recent studies by the World Bank [4] and the Organisation for Economic Co-operation and Development (OECD) [5] - to quote a few - pointed out the potential that robotics and artificial intelligence have to create more jobs. Some 85 million jobs may be lost, but at the same time, there is an expectation of 97 million new jobs that can be created worldwide [6].

A recent study in Japan revealed that an increase of 1 robot per 1000 workers would raise employment by 2.2% [7], corroborating other findings that robots and labour are complimentary. The critical aspect is how we manage the transition. But, as Guy Ryder recently said, "we must not fall into the trap of technological determinism". The way we frame the future of work will impact the result we want to have, "the future of work is the work of the future".

All countries stand to benefit from this technological revolution, but governments' adoption of enabling policies will play a key role. It is essential to facilitate business adoption of these technologies and avoid hampering tax and regulatory policies that impede innovation.

There are several ways robots can make our lives better and more efficient. They can take over tasks that are considered repetitive and even dangerous. For instance, they can ensure the safety of workers and employers. Workers also will have more time to dedicate to more exciting and creative tasks, like thinking, communication and problem-solving. Yet, enabling work transitions for new jobs demanding new skills will require appropriate policies and close dialogue with employers for the anticipated assessment of labour market needs.

Rising connectivity is creating new opportunities, with emerging technologies set to further the impact, such as artificial intelligence, robotics, autonomous vehicles, 3D printing, nanotechnology, biotechnology, or quantum computing. Small and medium enterprises with access to the internet experienced an estimated 11% productivity gain [8] (not just in developed countries but also in developing economies). This level will continue as long as internet access grows.

In the future, technological innovation will also lead to long-term gains in efficiency, productivity, and economic growth. As productivity rises, many more goods and services are produced, resulting in economic development.

Nonetheless, we shall not overlook the contingent risk of increased inequalities by creating digital divides between those who have access to technology and those who do not. Maximising the gains from today's digital era will require enabling environments for digital expansion and inclusiveness through appropriate policies.

According to UNCTAD's Technology and Innovation Report 2021, many countries, particularly the least developed, are unprepared to equitably use, adopt and adapt to the ongoing technological revolution. The report highlighted that "in developed countries in 2018, around 33% of the inhabitants had fixed broadband subscriptions, while in developing countries the proportion was only 11%". It went on to give more details: "in 2018, around 80% of people in Europe were using the internet, while in Sub-Saharan Africa the proportion was only 25% and in the least developed countries only 20%" [9]. Connectivity will remain a fundamental challenge to address in our efforts to leverage equally from the digital era.

Governments should address the digital divide by adopting policies that enable an environment of equal access to digital communication services, digital literacy and skilling opportunities, and equipment. Digital inclusion strategies will have to evolve and continuously adapt as technologies change and labour market needs evolve. For this to happen, close dialogue between governments and employers' and workers' organisations will be crucial for better assessing current labour market needs and challenges.

Technological skills need to ratchet up rapidly. Notably, a new mindset for constant learning could be more critical than ever. The significant shift from simple digitalisation to the Fourth Industrial Revolution has forced companies to re-examine the way they do business based on robust digital networks that easily cross borders. This revolution requires a fit-for-purpose workforce with the right skills.

Even before the crisis, employers found it very difficult to find talent. The International Organisation of Employers (IOE) and the International Labour Organization (ILO) published a survey entitled *Changing business and opportunities for employers and business organisations*. The survey showed that a large proportion of companies are looking for entirely different skills in recruits than three years ago. It also showed that the skills lifecycle is becoming shorter and demands rapid adaptation. Recruitment professionals explained that it is becoming more challenging to find people with the necessary skills.

Education systems need to reform to meet the needs of industry and the rapid changes in the world of work. We must do better on filling the skills gaps to develop employability in a world where technology is constantly changing. Most importantly, governments and the international community need to reverse the polarisation of opportunities and build a future where everyone can fit in with the current trends of the labour market.

Another example of technological tools driving change and triggering employment policy action is the rise in the use of platforms. The number of digital labour platforms worldwide has increased fivefold in the last ten years [10]. Indeed, the advancement in digital technologies has enabled easy access to goods, services, and jobs. Beyond the opportunities for filling employment gaps and boosting innovation and entrepreneurship via platforms, the platform economy as a whole is proving its great potential in the transition from informal to formal economies. The ease of accessing clients through digital technologies creates vast opportunities for growth from the informal economy. The digital nature of the transactions and the written conclusion of agreements and, in some cases, traceability are huge formalisation drivers.

In trying to assess the potential of platform work to incorporate more people into the labour market, we analysed data on labour market matching by temporary employment agencies. The numbers show that 35% of temporary employment agency recruits were either unemployed or inactive. We could assume that workers looking into opportunities offered within the platform economy could be alike. Over two-thirds of workers remain employed after the first assignment, with many transitioning to a structural relationship with the user-enterprise [11].

Another example of the positive aspects of leveraging platform work is the opportunities these bring to specific (underrepresented) groups of workers. While most platform workers appear to be male [12], it seems that women are successfully navigating e-commerce and other forms of participation via platforms. A recent study by the ILO highlights that not only has there been a rise in the share of workers or contractors using technology platforms, but that worldwide, four out of ten workers on online web-based platforms are women [13]. Female workers often highlight that platform work enables them to balance work and caring responsibilities. Whether driven by choice or compulsion in the face of restrictive social norms, women usually prefer to work from home [14]. Also, platforms can enable disabled workers to further participate in the labour market and improve their incomes. With the right policies, platform work can reduce the gaps in male and female labour force participation and inclusion of other underrepresented groups towards significant economic gains.

Also, entrepreneurs and start-ups are increasingly turning into digital platforms, whether by creating a new platform or by providing a product or service through an existing one. Platforms have turned into a space to foster innovation and entrepreneurship. Freelancers and independent workers have found in platform economy a vital tool. While platform owners do not necessarily own the product offerings by third parties or have direct control over their business partners resulting in employment relationships, in many cases, these technological means serve the filling of employment gaps through self-employment. Moreover, Online platforms offer some apparent benefits to SMEs. They provide a means to access new markets, sourcing channels, and many digital networks [15].

There is so much potential for AI, robotics, platforms, and digital processes to improve the quality of people's lives. All these are tools to help us improve working conditions and well-being, but not to replace us. We need to have the right mindset to embrace technology and adopt new ways of working.

Digital technology is a powerful tool to enhance work-life balance, productivity growth, income opportunities for many individuals if intelligently used and applied in our processes and practices. Still, it also raises personal disconnection challenges. Different and innovative answers in and outside the workplace are needed to avoid these counter-effects.

"We need to leverage the opportunities the digital era offers while also being committed to addressing the challenges it raises honestly and realistically".

The recent pandemic devastated many countries' economic and social systems, with jobs and business continuity at risk. Bringing about the necessary change is a shared responsibility between businesses, workers, consumers, and citizens. We must look at the future directly, not with fear, but with a sense of shared objectives and values. The important thing is to ensure that technology works for us and not the other way around.

The International Organisation of Employers is the largest network of the private sector in the world. We are strongly committed to social values. For more than 100 years, we have been representing business at the International Labour Organization and the UN and in different international forums and bodies. We have been intensively working to anticipate employment policy responses to shape the future of work before, during and long after the current pandemic.

References

1. Lund, S., Madgavkar, A., Manyika, J., McKinsey, S.S.: Company, what's next for remote work: an analysis of 2,000 tasks, 800 jobs, and nine countries, 23 November 2020
2. Praxis Global Alliance: Outpatient Healthcare Market in India, August 2021
3. McKinsey & Company: What 800 executives envision for the post-pandemic workforce, 23 September 2020
4. World Bank Group: The Future of Work: Race with—not against—the Machine, August 2018
5. The Future of Work OECD Employment Outlook (2019)
6. World Economic Forum Future of Jobs Report (2020)
7. VoxEU: Robots and employment: Evidence from Japan, 1978–2017, Daisuke Adachi, Daiji Kawaguchi, Yukiko Saito, 09 February 2021
8. Deloitte: Small and medium enterprises with Internet access have been found to have experienced an average of 11% of productivity gain, February 2014
9. UNCTAD: Technology and Innovation Report 2021, Catching technological waves (2021)
10. ILO: Digital platforms and the world of work in G20 countries: Status and Policy Action, June 2021
11. World Employment Confederation Economic Report 2021
12. Most studies find males make up 60 to 75% of platform workers (Pesole, Urzi Brancati, Fernandez Macias, Biagi, & Gonzalez Vazquez, 2018) (Berg, Furrer, Harmon, Rani, & Silberman, 2018) (ILO, 2021) (Edelman Intelligence/Upwork Inc., 2020) (McKinsey Global Institute, 2016) (Online Labour Observatory, 2021)
13. ILO: World Employment and Social Outlook: The role of digital labour platforms in transforming the world of work, Geneva, ILO (2021)
14. Berg, et al.: 2018, Sharma and Kunduri 2015, Betcherman 2017
15. OECD: SMEs in the online platform economy (2021)

Robotisation and Its Social Implications: The Skills Gaps and Digital Learning

Mª. Y. Sánchez-Urán Azaña(✉) and N. P. García Piñeiro

Labour Law and Social Security Law, Complutense University of Madrid, Madrid, Spain
{sanuran,npgarcia}@ucm.es, npgarcia@der.ucm.es

Abstract. This paper discusses the effects on the Labour Market in the "era of robots". Interaction and collaboration between humans and robots will increase. Employment and jobs will be transformed. The current model will have to adapt to its new stage of Digital Disruption.

1 Introduction

The vertiginous process of digitalization and robotization of the economy impacts on the Labour Market, so it is essential to analyze how this market could change in order to articulate some solutions under the premise of the principle of technological, social and legally responsible innovation. There will always be human work: "the relevant thing will be how the existing one is distributed and how to guarantee equal opportunities to have access to it" [1].

Technology changes the type of production and the types of jobs or workforce needed and will always require some reorganization of work and a new division of labour.

Technological innovation and division of labour influence each other: the former determines the latter; and the latter, when viewed in terms of worker specialisation, facilitates innovation and technological change (Eurofound 2018: 6).

In the context of robotisation, measuring its impact in terms of quantity of employment/quality or nature of work performed by humans has for years been the main concern. If we focus the analysis on the division of labour, and this from the employment rate/unemployment rate comparison, there is no consensus on the results and on the projections in the short, medium and long term.

How many times have we wondered how many existing jobs will be destroyed and how many new ones will be created because of AI and robotisation? Is there a fear, in the Anglo-Saxon expression automation anxiety, of being replaced by robots in the workplace? Perhaps we have not found an unequivocal answer because the effect of technologies is bivalent. Their potential for job destruction is counterbalanced by what appears to be job creation. As has been rightly expressed, "digitalisation destroys jobs because it replaces them, but it also creates jobs because it boosts growth, leads to the appearance of new goods and services, the emergence of new productive sectors and

Inbots, CSA, H 2020, G.A. n.780073 and FuWorkTech Spanish Proyect: *El impacto de la digitalización en las relaciones de trabajo, retos y oportunidades*, PID 2019-104287RB-100.

M. A. Grau Ruiz (Ed.): INBOTS 2021, BIOSYSROB 30, pp. 51–55, 2022.
https://doi.org/10.1007/978-3-031-04305-5_9

new occupations and tasks" [2]. Hence, the outcome of the interaction of the two effects is very difficult to predict. This would be the first and oldest scenario, that of a negative impact on employment, graphically described as technological unemployment (Keynes). However, other studies have observed [3] that those countries that have a higher density of robots per worker generally have a lower unemployment rate.

In a very different sense to job destruction, projections are made about the net positive impact of robotics on jobs and job quality. It warns - against the fears of some about the complete elimination of jobs as a result of automation - that, according to studies, only 5–10% of jobs will be completely replaced or substituted.

And the third, intermediate scenario, which seems the most realistic, labour market "polarization".

2 Labour Market Polarisation

The labour market polarisation has at least three different meanings. One, the oft-cited Frey and Osborne, who contributed to significantly fueling the debate when they published in 2013 the first version of their famous article "The Future of Employment" (Frey and Osborne, 2017), alarming about the percentage of jobs in the US that were at risk due to automation. Subsequently, the figure is qualified by the fact that there are many factors involved in the outcome, including regulatory aspects, political issues and social pressure, and they conclude that automation will mainly replace medium-wage jobs. Another distinguishes between high-skilled and unskilled jobs. And finally, one that differentiates between automatable and non-automatable jobs. While the second view assumes that technology will demand high-skilled workers and will not replace non-routine tasks and concrete tasks, the third shows how the latest technologies will also be able to displace such tasks performed by people with these medium skills.

In the scenario of "polarisation" of the labor market, it is said that the substitution effect will not be so much of jobs but of specific tasks, particularly those that are considered repetitive, whether physical or data processing. In this regard, it is argued that the type of tasks that robots or other automation tools will not be able to perform soon, even with advances in artificial intelligence, are those that focus on creativity, empathy, persuasion, an understanding of knowledge, and a high level of sensory-motor skills, which are influenced by the level of education or training of workers. But paradoxically, it has also been found that technology can contribute to creating jobs where intermediate digital skills are required and these can be more easily replaced in a technological advance.

In this context of uncertainty about job creation, reduction or polarisation, it is necessary and urgent to take strong measures in the labour market. Even if it is claimed that there will be no negative impact on jobs, it is a different matter whether the labour market has workers with the necessary skills for the new jobs that may be created. In this respect it has been rightly stated that "countries that are able to train their workers in the new skills and retain talented workers can shorten the transition period. This will allow them to be more competitive and take advantage of the benefits of automation [3].

In order to face the future of work and employment successfully, it is essential to adopt measures to redirect the division of labour in the context of the training and qualification of workers in order to tackle skill shortages, skill mismatches, skills gaps and the lack of lifelong learning.

2.1 Skill Gaps and Wage Gaps

The "competence" gaps (the so-called skill gaps) are another factor in the wage gaps. If the worker has average competences/skills and his job can be robotized, he will not be able to aspire to a job suited to his training, he will have to do lower qualified work and, therefore, suffer a reduction in his salary. If the worker does not have these required skills, the gap will occur in the short term and, therefore, a new type of inequality or vulnerability will be created in the labour market derived from the difference or mismatch between the skills that employers are looking for and the skills that workers have. Observing these inequalities in the context of the wage gap, the process of robotization and the advance of technological innovation may generate greater polarisation because the increase in the wages of low -and medium- skilled workers. So more low -skilled jobs will be available with fewer opportunities for wage growth [4], perpetuating the problem.

2.2 Training and Qualification: Digital Learning

Training for a qualification its necessary to address the gaps and deficiencies detected in the design and immediate implementation of specific measures. A public-private collaboration programme should be advocated that precisely sets the objectives, intermediate and immediate, and their different phases. Today more than ever, the digital transformation requires workers with multipurpose training and lifelong learning to adapt to the technological advances that are taking place.

It's necessary to train in digital skills and support employment through access to, among other measures, digital learning. It is estimated that up to 90% of jobs in the EU will require some form of digital competence in the coming years [5] (That is why the European Commission is firmly committed to the promotion of advanced digital skills as one of the five priority areas of the Digital Europe 2021–2027 Programme. But also, the necessity to qualify and train in other competences or skills, in particular those that can be qualified as "social and emotional skills", among them, advanced communication and negotiation skills; interpersonal skills and empathy; leadership and management; entrepreneurship and pro-activity; adaptability and continuous learning; teaching and training should not be overlooked.

International Organisations and EU institutions (within the framework of the European Pillar of Social Rights) have been focusing on lifelong learning as a guarantee of people's employability for a long time. In this transition period, the question is who is to pay for the re-skilling and retraining of those who are currently in the workplace but may lose their jobs as a result of robotization, and how.

The latest Spanish labour reforms, especially in 2011 and 2012, enshrined important advances in vocational training for employment in the individual employment relationship that allow workers to face the permanent challenge of training demanded by the digital transformation, among others, the training account [6]. Some of these are in line with the measures proposed by the European Commission's High-Level Expert Group on the Impact of the Digital Transformation on EU Labour Markets (2019).

Training in digitization and robotics will certainly help to minimize the impact of the digital transformation on job destruction and skills mismatch. Moreover, the establishment of an effective vocational training system that allows workers to be constantly retrained, to fully develop their skills, to facilitate their job transitions and to enable companies to have a skilled workforce with which to respond to technological change and robotization, must be an essential component of employment policies.

2.3 Proposals of Vocational Training for Employment

The current situation advises a calm reflection on the future of vocational training for employment and, in the context of digitalization in general, and robotisation, proposals can be made in this regard. Among them: inclusion of a new priority group in the Employment Act, that of workers with a deficit of training in new technologies and digitization and with difficulties of integration into the labour market [7]. The implementation of employment stimulus measures for certain productive sectors or geographical areas that may have been affected by the eruption of new technologies. The establishment of guidance and continuous professional training actions in the field of digitalization and new technologies as an adjustment mechanism to avoid the negative effects on workers, especially older and less qualified ones. The promotion of entrepreneurship and self-employment as an option for labour insertion, to meet the enormous opportunities for growth and employment made possible by new technologies. And finally, assessing the bonuses or reductions in social contributions for companies that introduce new technologies, maintain the level of employment and qualify their workers [8]. The aim of the measure is to contribute to the retraining an outplacement of these workers from within the company, without expelling them from the labour market as far as possible. The incentive to technological innovation and the conversion of companies and activities from the traditional model to a more technological model must be accompanied, at least during the transition phase, by economic incentives for those companies that invest really and effectively in the technological "training" of their workers.

3 Conclusion

It is urgent to prioritise the modernisation of the vocational training system for employment. It must be adapted to the new needs demanded by the labour market derived from digitalisation. The ultimate aim must be to avoid skills gaps, as it will allow to match labour supply and demand in an increasingly robotised labour market.

References

1. Sánchez-Urán Azaña, Mª.Y., Grau Ruiz, Mª.A.: Plataformas Digitales y Robótica: su proyección en el Sistema de Seguridad Social, p. 21. http://eprints.ucm.es/id/eprint/64676/
2. Pérez García, F., et al.: Cambios tecnológicos, trabajo y actividad empresarial: el impacto socioeconómico de la economía digital, CES, Resumen ejecutivo. http://www.ces.es/documents/10180/5228908/2020

3. López Sánchez, I., Grau Ruiz, A., Sánchez-Urán Azaña, Mª.Y.: The impact of robotics and computerization on the labour market: inclusive insight from a Law and Economics perspective. Revista Innovacion y Derecho Digital **3**, 19 (2019)
4. Taes, S.: Robotisation and labour law: the dark factory: the dark side of work? In: Artificial Intelligence and the Law, p. 295. Cambridge, Antwerp, Chicago (2021)
5. European Commission report Digital Skills. New professions, New educational methods, New Jobs (2019)
6. Aguilera Izquierdo, R., García Piñeiro, N.P.: La formación profesional para el empleo: reflexiones para su reforma. Revista Española de Derecho de Trabajo **147**, 485 (2010)
7. García Piñeiro, N.P.: Política de empleo y nuevas tecnologías. Derecho del Trabajo y Nuevas Tecnologías, Tirant lo Blanch, pp. 183–184 (2020)
8. Díaz, A., Grau Ruiz, Mª.A.: Social security and robotization: possible ways to finance human reskilling and promote employment. Paladyn J. Behav. Robot. **11**(1). https://www.degruyter.com/document/doi/10.1515/pjbr-2020-0020/html

A Skilled Workforce – Ready to Contribute to Tomorrow's World of Work

Maarten Goos[1,2](✉)

[1] Utrecht University School of Economics, Utrecht, The Netherlands
M.Goos@uu.nl
[2] Instituut Gak, Hilversum, The Netherlands

Abstract. The digital transformation is rapidly changing the demand for workers' skills, and this poses several challenges. One such challenge is how do we define 'skills' in the digital economy. Another challenge is who invest in workers' digital skills? This chapter discusses policies to better measure digital skills, personal learning accounts; 2) policies focussed on training of workers; and 3) the role of intermediaries to help reduce the skills gap.

1 Introduction

The digital transformation is contributing to skill mismatch and shortages that require investments in education and employee training. Considering these challenges, several actors, including the European Centre for the Development of Vocational Training, the European Commission, the Organisation for Economic Cooperation and Development (OECD) and European Union Member States, have focussed on the question of how to achieve a better alignment of skill supply and demand.

One important question is how to define 'skills' in the digital economy. Although the various existing measures of worker skills are informative about shortages, these measures usually do not account for workers' task experiences and soft skills as well as their relevance for individual employers, each with their specific workplaces and related skill needs. Another important question is whether workers, employers, some third parties or a combination of these should invest in workers' digital skills?

Given these questions, this chapter discusses policies to better measure digital skills, personal learning accounts; 2) policies focussed on training of workers; and 3) the role of intermediaries to help reduce the skills gap.

The digital transformation is rapidly changing the demand for workers' skills and task competencies. This way, the digital transformation is contributing to skill mismatch and shortages that require investments in employee training. In light of these challenges, several actors, including the European Centre for the Development of Vocational Training, the European Commission, the Organisation for Economic Cooperation and Development (OECD) and European Union Member States, have focussed on the question of how to achieve a better alignment of skill supply and demand, with a focus on: i) understanding how countries collect and use information on skill needs; ii) investigating cost-effective training and labour market policies to tackle skill mismatch and

M. A. Grau Ruiz (Ed.): INBOTS 2021, BIOSYSROB 30, pp. 56–58, 2022.
https://doi.org/10.1007/978-3-031-04305-5_10

shortages; iii) studying the incentives of training providers and participants to respond to changing skill needs; iv) setting up a database of skill needs[1].

Despite the extensive focus on collecting survey data to measure different versions of skill gaps, evaluating existing training policies and collecting survey data, several important challenges remain. One such challenge is how do we define 'skills' in the digital economy.

One definition of a worker's skills is the formal education that a worker received during full-time education before entering the labour market. Given this definition of skill, one can then think of how technological progress has changed the demand for workers with more relative to less full-time education[2].

However, a more precise view would be that digitalisation is changing the demand for tasks that workers do on-the-job because some tasks can be automated, but others cannot. Consequently, digitalisation will change the demand for workers with different levels of formal education only indirectly through changes in on-the-job task requirements for workers[3].

This decoupling between workers' formal education levels and their task competencies poses the question of how to define skills when thinking about the impact of the digital transformation on labour markets[4]. In part due to this decoupling, many different classifications of skills (e.g. years of schooling, occupational or sector experience, tasks done in an occupation and soft skills such as personality traits) have been developed and are used to measure skill gaps[5].

Although the various existing measures of skill gaps are informative about skill shortages and abundancies for the labour market, they are less accurate regarding individual workers, each with their specific formal education, task experiences and soft skills as well as for individual employers, each with their specific workplaces and related skill needs.

One solution to this problem would be to use big data and machine learning techniques to better inform individual workers about their specific skill set and help them find jobs that better match their skill set. Similarly, such a tool could be used to help employers find workers that better fit their skill requirements, thereby reducing skill gaps.

[1] OECD, Getting Skills Right: Assessing and Anticipating Changing Skill Needs", Chapter 4, 2016.

[2] This was the consensus thinking in the academic literature until recently. See, for example, Goldin and Katz, The Race Between Education and Technology, Harvard University Press, 2009.

[3] Acemoglu, D. and Autor, D., Skills, Tasks and Technologies: Implications for Employment and Earnings, Handbook of Labor Economics, 2011, Volume 4b, https://economics.mit.edu/files/5571.

[4] Goos, M., Manning, A. and Salomons, A., Explaining Job Polarization: Routine-Biased Technological Change and Offshoring, American Economic Review, 2014, 104(8), pp.2509–2526.

[5] For an example of such a tool, see https://www.media.mit.edu/posts/how-skills-affect-your-job-trajectory-and-their-implications-for-automation-by-ai/, although this algorithm is not (yet) programmed to reduce skill gaps.

Additional policy recommendations are 1) digital skills personal learning accounts; policies focused on the delivery of training, career guidance services and quality assurance; and 3) policies to support intermediaries to help reduce skill gaps:

1. Enabling **digital skills personal learning accounts**, allowing workers to acquire relevant skills throughout their careers in order to stay relevant in rapidly transforming, digital labour markets. The accounts would belong to the worker and would be portable from job to job. Details such as contributions, number of hours per year, top ups, eligible expenses, withdrawal processes and taxing schemes will be determined later.
2. Scaling up **career counselling** and creating **innovative learning environments** to enable better career choices and active pursuit of relevant training for all Europeans. Career counselling could be supported through establishing quality training standards and "digital literacy" for career counsellors at the European level. Communities of practice could foster informal group learning at workplace.
3. Supporting **labour market intermediaries to reduce structural skill gaps especially for women in Science, Technology, Engineering and Mathematics (STEM), workers at risk of automation and the low-skilled**. Intermediaries (such as public employment services, outplacement offices or temporary agencies) would invest in on- the-job training provided they can recoup the training cost from employers who on their turn will benefit from trained workers.

Inclusive Robotics for the Employment of People with Disabilities

C. De Fuentes Gª-Romero de Tejada[1](✉) and C. Juanes Calvo[2,3]

[1] Universidad Complutense Madrid, Madrid, Spain
carlosdefuentes@pdi.ucm.es, carlosdf@ucm.es
[2] Learning and Development BDGS, Ericsson, Madrid, Spain
cjuanes@faculty.ie.edu
[3] IE University, Madrid, Spain

Abstract. Robotics can be a great ally for the inclusion of people with disabilities in the labor market. Job adaptations that previously seemed unthinkable are now possible thanks to technology. In this work, several examples of technological innovations that can facilitate the inclusion of this cohort are reviewed, especially for physical and sensory disabilities. Inclusive robotics, thanks to the opportunity that Artificial Intelligence creates through sensors, open the opportunity of wider accessibility for people with disabilities, promoting their independent life and participation in all aspects of life on equal terms with the rest of the population.

1 Introduction

According to the World Economic Forum's Future of Jobs 2020 report, the impact of automation, robotics and artificial intelligence on the world of work will be very significant in two directions. On the one hand, a significant percentage of current jobs will be destroyed, affecting above all people with fewer skills and who perform more basic activities. On the other hand, it may allow for the creation of new jobs mainly in sectors linked to human care, technology industries and content creation [1].

Regardless of the number of jobs it is expected to destroy or create, the fact is that the social footprint of this new Industrial Revolution will be evident [2]. People who don't have a qualified level of education and sufficient technological skills will find it difficult to enter or remain in the job [3, 4]. This will have a particular impact on those at risk of exclusion. However, technology can also be an ally to achieve accessibility to jobs that were unthinkable until now. This is the approach we are going to follow in this article with the so-called inclusive robotics.

The objective of this communication is to investigate the positive impact that robotics can have on a wider labor inclusion of people with disabilities (PwD) thanks to improvements in the adaptation of jobs.

We have structured the analysis and content as follows: first, we will review the current situation of PwD in the workplace and the main international standards that promote

Article framed in the research project "FuWorkTech. El impacto de la digitalización en las relaciones de trabajo: retos y oportunidades". (Ministry of Science and Innovation: PID2019-104287RB-100; R+D+I projects).

M. A. Grau Ruiz (Ed.): INBOTS 2021, BIOSYSROB 30, pp. 59–65, 2022.
https://doi.org/10.1007/978-3-031-04305-5_11

universal accessibility, with special emphasis on the Spanish situation. Second, we will analyze the definition of the term robotics, and how it could address the opportunity of inclusive robotics. Third, we will review some examples of technology for job adaptation. We have classified them according to the type of disability. We will finalize the paper with some conclusions and recommendations.

2 Current Work Situation of People with Disabilities

Over a billion people in the world and 100 million people in the EU live with a disability. In 2018, the employment rate for PwD (aged 20 to 64) was 50.8% vs. 75% for people without disabilities. Besides, in 2018, 22.6% of PwD younger than 60 years were living in households with very low work intensity vs. 7.1% of people without disabilities [5].

On December 13th, 2006, the United Nations General Assembly approved the United Nations Convention on the Rights of Persons with Disabilities (UN CRPD). Article 9 of the Convention sets out general obligations for Member States to ensure that PwD have access on equal bases to the physical environment, transportation, information and communication, including information and communication technologies and systems (ICTs). Accessibility is also identified in Article 3 (f) of the Convention as one of its eight principles.

In Spain, the General Law on the rights of PwD and their social inclusion (Royal Legislative Decree 1/2013, of November 29) establishes universal accessibility as a fundamental principle, which naturally encompasses work activity. It is developed, among others, in articles 22, 35 and 40, especially, in the second section of this last article.

In Spain in 2020, according to the latest published statistics [6] almost two million people, 6.3% of the active population, had a disability. The most relevant fact for this analysis is that only 34% of them were active, i.e. they were working or looking for a job. Therefore, the vast majority of PwD in our country were totally distant from the world of work, 40% less than the rest of the population. There are many factors that influence this situation, but the lack of accessibility for PwD to perform a job is undoubtedly an essential aspect.

Taking into account only those who are active, 22.23% are unemployed, almost seven points more than the non-disabled population. Only 26.7% of working age PwD are employed, compared to 64.3% of the general population. In particular, the rates of inactivity and unemployment are centred on people with intellectual disabilities and mental health problems, the two groups that, as we shall see, benefit the least from the technological adaptations that we are going to discuss in this document.

In short, although the trend in recent years is good because almost 180,000 people have been incorporated in the last 10 years, the truth is that there is still a lot of work to be done.

3 Robot Definition

There are different conceptual approaches to Robotics and robots. In this work, we start from a broad vision and embrace the definition of the British Encyclopedia, as "any automatically operated machine that replaces human power, although it does not resemble human beings in appearance or perform its functions in the same way" [7]. The essential aspect is that it is "a mechanism guided by automatic controls" [8] and, in this way, it encompasses inclusive robotics (predominantly interactive in services sector and others, i.e., healthcare) [9]. We also include the technology behind the actual robots, which nowadays is mostly sensory capabilities thanks to Artificial Intelligence (AI) capabilities.

It would be advisable for this vision of inclusive robotics to have a greater presence in the regulatory proposals being put forward by the European Union. Its positive implications, especially for the PwD, would require a boost from European regulations. In this line, a noteworthy example can be found in the European Parliament Resolution of 20 October 2020, Framework on the ethical aspects of artificial intelligence, robotics and related technologies [10], paragraph 50 of which states:

"Considers that artificial intelligence can help to make better use of the skills and competences of people with disabilities and that the application of artificial intelligence in the workplace can contribute to inclusive labour markets and higher employment rates for people with disabilities".

4 Some Examples of Inclusive Robotics for People with Disabilities

4.1 Physical Disability [11]

There have been advances in the handling of computers, especially the movement of the cursor, through for example the **HeadMouse**. Through this system it is possible to control its movement with slight movements of the head and perform click actions with facial gestures. These movements are captured by a camera installed on the computer. Another advance is **VirtualKeyBoard**, a keyboard that appears on the screen and that PwD can operate through the HeadMouse.

But the most innovative system in this area is the eye- tracking system, where control is done exclusively by the eyes. The computer stops where the eye looks by emitting infrared rays that identify the eye's pupil. There is no need to click, because a side panel shows the various possible functions that are also selected with the eye.

We also refer to the opportunity of Personal Robots, such as **PRA Project** (Personal Robotic Assistant). Most of them refer to humanoid robots that can move easily in a work environment and that serves for the presence of the teleworker. These robot can be controlled by a smartphone or tablet, or by voice control. It is in testing phase.

4.2 Sensory Disability: Speech Impairment

Speech Assistant AAC [12] is a text speech app designed for people who have speech impairment. This may be in the case of aphasia, MND/ALS, autism, stroke, vocal cord problems or other speech problems.

With the app you can create categories and phrases, which are placed on buttons. With these buttons you can create messages that can be shown or spoken (text-to-speech). It is also possible to type any text using the keyboard.

4.3 Sensory Disability: Blind People

There are four different technologies that are used to overcome the barriers of blind people.

1) ***Smartwatch*** [13]: Unlike most assistive smartwatches for the blind which rely on audio prompts, the Dot displays messages four braille characters at a time on its screen. Its round face displays four cells of six balls each, and allows users to send simple replies or actions back through its two side buttons (as shown in Fig. 1).

Fig. 1. The Dot braille smart watch.

2) ***Smart glasses*** [14]: Inside are two OLED screens and powerful prisms. There's also a processor in the headset and it's all connected to a small controller/battery pack (it gets 3-to-4 h of battery life). Images collected by the camera and data from the sensors, which help keep the image in focus, is all processed in the visor and then displayed on tiny dual screens just millimetres from person's eyes *(as shown in* Fig. 2*)*.

Fig. 2. Smart glasses are digital eyes for the legally blind.

3) ***Seeing AI App*** [15]: The app not only recognizes a person based on his or her face, but can also relay mood based on facial expressions, estimate a person's proximity

to the user, and describe physical characteristics. This app can also identify products based on their barcode, a feature aimed at making shopping easier. The app emits a series of beeps to indicate how close the viewfinder is to the barcode, helping users align the camera properly. Similarly, when reading documents to users, Seeing AI will provide spoken hints to ensure that all corners of the paper are properly captured *(as shown in* Fig. 3*).*

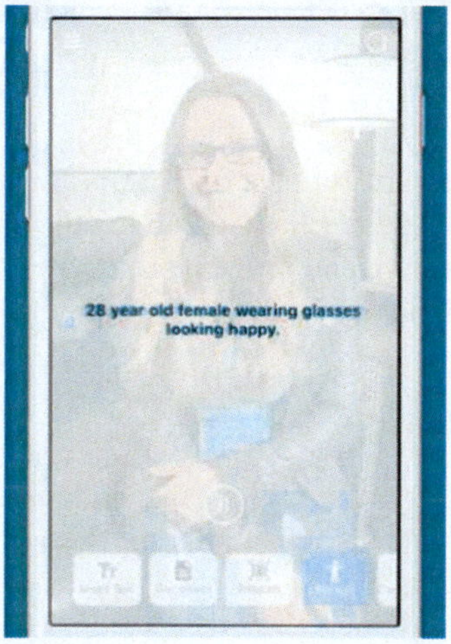

Fig. 3. This app makes life easier for visually impaired.

4) ***Edgguide*** [16]: is a collaboration between CACI [11] and the Blind Institute of Technology (BIT) to provide an indoor micro-localization solution for the blind and visually impaired (BVI) so everyone can enjoy a technology enhanced experience. In addition, the Denver Museum of Nature & Science (DMNS) hosted a six- month pilot of the capability ending in the spring of 2020. By leveraging this technology, the museum was able to expand their services to a broader community: precision navigation for the BVI community, a 'find your buddy' feature, curated audio content that augments the tactile experience of blind and visually impaired, and exhibit-specific audio content available in multiple languages for audio learners.

4.4 Sensory and Intellectual Disability

1) ***The PACA robot*** [17] is a personal and accessible assistant. Its function is to guide and inform and it can be a support for the labour inclusion of PwD. It is a humanoid robot that measures just over one metre and has the option to be programmed and designed to interact and assist people. The robotic device provides information in an accessible way for deaf, blind, visually impaired, intellectually disabled and elderly people. Its technology is able to recognise both verbal and non-verbal language. People can ask it questions orally or via text by writing on the screen on its chest. Thanks to this screen, it can communicate with deaf and dumb people *(as shown in* Fig. 4*).*

Fig. 4. The PACA robot is a personal and accessible assistant.

2) ***Affectiva*** [18]: this is a research project from Stanford University, which grew out of the MIT Media Lab, and Autism Glass, involving the use of artificial intelligence to automate emotion recognition and provide social cues to help people on the autism spectrum to interact in social settings.

4.5 For All Types of Disability

Inclusive employment, where robots and humans work as a team [19]. This is a restaurant (Dawn Cafe in Tokyo) where a humanoid robot serves customers but is controlled remotely by workers, including PwD. This is a project to create employment opportunities for workers with disabilities and other needs. The robots are controlled from home by PwD. There are about 20 robots that move around the restaurant space to serve customers *(as shown in* Fig. 5*).*

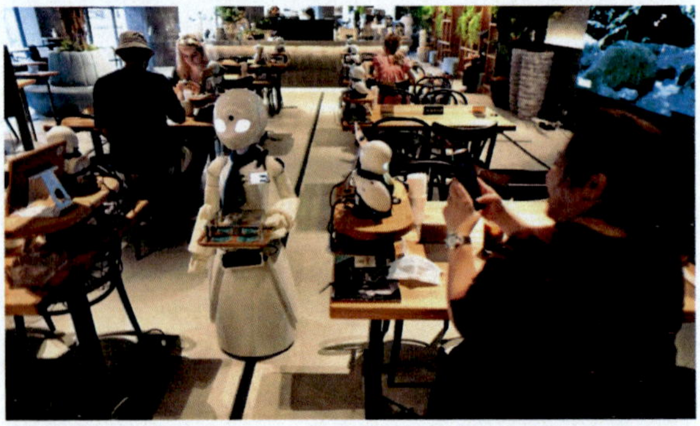

Fig. 5. Image of Down Cafe in Tokyo.

5 Conclusion

Technological advances in the area of inclusive robotics are very encouraging. Existing social barriers to jobs can be overcome thanks to technology. The examples that have been seen in this work allow PwD to carry out their work. Therefore, it can be concluded that inclusive robotics supports the labor inclusion of this group of people. However, accommodations are rare and furthermore only address barriers to physical and sensory disabilities. Cognitive accessibility is left out, whose most common tools are easy reading and The Augmentative and Alternative Communication Systems [19]. There is no progress being made in adaptations for psychosocial disabilities neither.

References

1. World Economic Forum: The Future of Jobs Report 2020. https://www.weforum.org/reports/the-future-of-jobs-report-2020
2. INBOTS: INBOTS White Paper on Interactive Robotics Legal, Ethics & Socioeconomic Aspects, p. 76 and following. http://inbots.eu/wp-content/uploads/2021/09/D.2.2-FINAL-INBOTS-White-Paper-WP2_-Legal-ethical-socioeconomic-aspects-.pdf
3. Kimbrough, K., Blue, A.: Your next change of job could be easier than you think. https://es.weforum.org/agenda/2020/10/tu-proximo-cambio-de-trabajo-podria-ser-mas-facil-de-lo-que-piensas/
4. Sánchez-Urán Azaña, Y., García Piñeiro, N.P.: Robotisation and its social implications: the skills gaps and digital learning. In: Inbots Conference (2021). https://eprints.ucm.es/id/eprint/66952/
5. International Labor Organization: Global Business and Disability Network and Fundación ONCE, An inclusive digital economy for people with disabilities, p. 8 (2021). http://www.businessanddisability.org/wp-content/uploads/2021/02/inclusiveDigitalEconomy.pdf
6. National Statistical Institute: Employment of people with disabilities, 2020, December 2021. https://www.ine.es/prensa/epd_2020.pdf
7. Mercader Uguina, J.: La robotización y el futuro del trabajo. Revista Trabajo y Derecho: nueva revista de actualidad y relaciones laborales **27**, 13–24 (2017)
8. Ortega, A.: La imparable marcha de los robots, Madrid, Alianza Editorial, pp. 14–16 (2016)
9. Sánchez-Urán Azaña, Y., Grau Ruiz, M.A.: Robotics and work: labor and tax regulatory framework. In: International Congress Technological Innovation and Future of Work, Santiago de Compostela, 5th and 6th April 2018, pp. 5–6 (2018). https://eprints.ucm.es/id/eprint/47718/
10. European Parliament resolution of 20 October 2020 with recommendations to the Commission on a framework of ethical aspects of artificial intelligence, robotics and related technologies (2020/2012(INL)), P9_TA(2020)0275. https://www.europarl.europa.eu/doceo/document/TA-9-2020-0275_ES.html
11. Cordero Gordillo, V., et al.: La integración laboral de personas con discapacidad en la era digital. Perspectiva internacional. In: ILO Interuniversity Congress on the Future of Work, Vol. II, Ministry of Labor and Social Economy, pp. 325–326 (2020)
12. https://www.iaccessibility.com/apps/speech/index.cgi/product?ID=330
13. https://mashable.com/2017/02/22/dot-smartwatch-retails/?europe=true#bgBU0sva5aq1
14. https://mashable.com/2017/02/15/esight-3-smart-glasses-vision-legally-blind/?europe=true
15. https://time.com/4854658/iphone-apps-visually-impaired-microsoft/
16. https://blindinstituteoftechnology.org/edgguide/
17. https://www.discapnet.es/la-robotica-asistencial-innovacion
18. McKinsey Global Institute: Report «Notes from the AI frontier: applying AI for social good», Noviembre 2018. https://www.mckinsey.com/featured-insights/artificial-intelligence/applying-artificial-intelligence-for-social-good
19. https://www.clarin.com/fotogalerias/empleo-inclusivo-robots-humanos-trabajan-mano-gastronomia-japon_5_O6j6GqrZ_.html

Diligence of IT Security: Employee Awareness Tests as a Legal Challenge in Portugal

David Carvalho Martins, Gonçalo Asper Caro, and Tiago Sequeira Mousinho(✉)

Faculty of Law, University of Lisbon, Lisbon, Portugal
tsm@dcmlittler.com

Abstract. This paper aims at the legal challenges provoked by the awareness tests on IT security employees. The new era of *Robotics*, *Artificial Intelligence (AI)*, *Deepaking* and *Computer Hacking* raised suspicions and lead the Market to search those capable of (somehow) minimize risks of digital assaults or espionage-like acts. The IT security personnel is subject to equal treatment and can be subject to the employer's (right to) control. Therefore, we address the origins of such thematic, the consequences (and admissibility?) of controlling IT security employees. Is the Portuguese jurisdiction ready and set for the challenge? Where do privacy provisions fall?

1 Introduction

The age of Labour or Industry 4.0. rather seems like something from the past. The rise of teleworking or remote working, alongside the Covid-19 pandemic virus, has tested traditional rules by storm (as well as the emergency measures taken "at the last minute"). Digital and intellectual forms of work have acquired a whole new meaning, where corporate and personal data – most importantly secrets – are endangered. Labour & Employment Law is currently facing a silent (yet dangerous) challenge on the application-interpretation of protective norms [1]. The historic North-American judge Richard Posner has once stated, around the 1980's, that privacy needs protection because there is always someone trying to access and pry on [2].

James Delaney, main character of the Taboo TV series said in such show: "*secrets are weapons to me*", back in the times of the Pacific trading, marked by the expeditions (and expansion) of East India Company, done by British subjects (late XVIII century). As the saying goes: "secrets are the business's soul". The same expression can be used to characterize today's Industry 5.0., whether to compete, market agents need to established safe and private parameters (both *physical and digital*) in order to: (i) protect their secrets; and (ii) to stay relevant in the *Market of Ideas, Knowledge and Know-how*. One may take for the example the famous cases of F1 racing teams (*engines, models, designs, among others*).

Taking into consideration the year of 2021 and 2022, Portugal firms have been targeted by several (and heavy) cyberattacks (Portugal is the 2nd country in the world most affected by spam and phishing, according to Safe Communities Portugal) – bringing concern to all employers on personal data protection and trade secrets; and to employees as well, since they are the first line of protection, or the "gatekeepers".

M. A. Grau Ruiz (Ed.): INBOTS 2021, BIOSYSROB 30, pp. 66–70, 2022.
https://doi.org/10.1007/978-3-031-04305-5_12

Alongside the rise of Industry 5.0. we have witnessed closely the ascension of new professions and services to make ends meet. The curve of demand has produced or crafted a new line of offers. The "digital shipments" have found an answer to secure their route safely, counting with the participation of IT security and their awareness.

Deepfaking, phishing, usage of trojans, spyware and malware have been pointed as key tools for virtual espionage (and privacy torts) as well as corporate data breaches (commercially sensitive data or trade secret breaches). Long goes the time where espionage was effectuated directly by human agents, entering by their own feet in the physical parameters of the company to spy (and pry?) on.

There have been complaints where the employee is "tricked" by a third party that acts, talks and seems like the "real employer". Imagine that an employee receives a Whatsapp message (with deepfakes, whether audio, text or a video) where the (false) "employer" demands determined file of clients' rap sheet. What if the *employee, supposedly complying with his duties, breaches such data*? What if an employee receives a rather suspicious e-mail – a simple and typical case of mail phishing – and opens it out of curiosity and, all the sudden, *all of his computer information (personal and professional) is collected by an unknown person*? [3].

Employers and employees (IT security) mutually collaborate to face the risk of breach, virtual attacks and sophisticated "hacking" phenomenon. There are threats of several orders (shapes or forms): (i) to obtain (illicitly) information of the company, to compete or ascertain a determined position in the market; (ii) to simply steal company data and sell to (other) possible interested market agents (mercenary-like acts); and (iii) to simply pry on.

Employers need to be ensured that all employees comply with their action, that they fulfil their part of the domino structure, in order to sustain the professional activity in an aggressive and competitive market. But begets the question, now: how does one, exactly, measures the awareness demanded to protect personal and corporate data? In other words, is it possible for the employer to test the digital parameters and safety premises of the company that he controls? By testing the awareness of the employees responsible for such tasks?

Awareness tests – *provoked by the employer* – have been studied has a potential offense to personality rights or as a privacy intrusion of employees, mainly IT security personnel [4]. In Portugal, for example, it seems to be forbidden to establish the so called "*honesty tests*" (for example, leaving valuable objects "forgotten" in order to test the ethic morals of employees, or even their loyalty) [5, 6] and it is also forbidden for the employer to set/create traps or artificial schemes that are apt to lead the employee to misconduct (as an expression of *venire contra factum proprium*, the abusive exercise of a right, for being contradictory) [7].

On the other hand, however, Superior Courts (*Supremo Tribunal de Justiça* [8] *and Tribunal da Relação de Lisboa* [9, 10]) have accepted the intervention of "*mystery clients*", contracted by the employer to obtain certain reports of his professional service (among the elements, the employee's awareness).

Certainly, it is not easy to provide a consensual or a unanimous response. Personality rights, namely the employee's privacy (at home, while teleworking, or at the company) is a rather sensitive thematic. If one it to accept the usage of such tests, it is important to note that the risk of offensive (privacy) intrusion is highly increased, especially in the context of the employment contract (*per si* highly fertile in privacy offenses).

A limit is to be established – the employee cannot be considered a commercial object, nor a machine or a program. At the end of the day, the employer has a duty to ensure a safe workplace environment (whether physical or digital, free from assaults and entrapments).

But if one is to refuse such measure, other (several) consequences can be pointed, such as the inability of the employer to scrutinize the work performed, the equal inability to test the diligence and attention due in these positions of work, and the total incapacity of the employer to face "surprise challenges". We must note that the risk of breach – attacks from third parties, *hackers* and curious hunters – absorbs all kind of information (personal data of employees included, not only corporate sensitive information).

Reaching this point, we believe that IT (employee) awareness procedures might be implemented: (i) on a vocational training level ("better to prevent" as it is said), establishing a first safeguard based on education and culture for the protection of internal information – initial formations might also be an alternative "to ensure an employee's familiarity with the company's security procedures when employment commences" [11]; (ii) through periodic reviews of acquired knowledge on such vocational training, while periodically remind employees on confidentiality and trade secret protection, through written documents (e.g., internal regulations or individual non-disclosure agreements (NDAs) to operate during the contract (the so called "reminder") and after termination (*post finitum* effect); (iii) to proceed with reasonable and loyal testing of awareness, taking into account the level of detail (and ingenuity) in such threats – since employees can be "significant contributors to the loss of intellectual assets" [12].

We believe that whistleblowing [13–17] systems (see the Portuguese legislation on this matter [18]) can articulate the compliance method for the protection of internal information. Delimiting, precisely and reasonably, the content of the information to be disclosed externally (strictly necessary to make the report effective and to ensure the protection of the whistleblower). This could be useful for accountability purposes for the workers themselves, as well as to prevent insider attacks. It also promotes compliance with the legal obligation to create internal reporting channels.

Therefore: (i) awareness testing can be settled and (ii) employees can be evaluated – consequently (and eventually) sanctioned based on the Portuguese Labour Code [19]. However, this type of analysis depends on objective considerations (of the normal execution of the professional activity and the abstract-type of employees) and subjective considerations, in regard the concrete employee (such as information that he carries, qualities, experience, age and vocational training provided) – in light of a *bonus pater familias* criteria [20–22].

2 A Few Conclusions

We believe that there is a possibility to establish or trail somewhat of a middle term. Certain forms of control can be operated and are legal (articles 20 and 21 of the Portuguese Labour Code) when they are reasonable, non-abusive and necessary (article 18, number 1 and 2 of the Portuguese Constitution).

Lack of awareness is apt to *cost everything* and break the chain or domino structure of one's company; but the application of such legal fundaments cannot be used to *harm the employee.*

Establishing a balance, by having both interests at presence, is important to determine a safe workplace environment for both the employer and the employee. Even on the edge of Industry or Labour 5.0.

References

1. Steely, R.P.: Stealing home: the quiet game of trade secret theft in the teleworking economy. The Natl. Law **X**(293) (2020). https://www.natlawreview.com/article/stealing-home-quiet-game-trade-secret-theft-telework-economy. Accessed 29 Apr 2021
2. Posner, R.: The Economics of Justice, 1st edn, pp. 231–310. Harvard University Press (1981)
3. These examples were given by Rui Shantilal at a conference, in a brief exercise that we have borrowed for the present paper. Rui Shantilal, "deepfakes", In: "Conferência sobre os 18 meses de RGPD", held by the Portuguese Catholic University of Lisbon, School of Law, 05 December 2019
4. Boehm, F., Hey, T., Ortner, R.: How to measure IT security awareness of employees: a comparison to e-mail surveillance at the workplace. EJLT **7**(1) (2016). https://ejlt.org/index.php/ejlt/article/view/500. Accessed 29 Apr 2021
5. Gomes, J., do Trabalho, D.: Relações individuais de trabalho. Coimbra Editora, vol. I, pp. 311–336 (2007)
6. Mousinho, T.S.: Detetives em contexto laboral. Prontuário de Direito do Trabalho, CEJ, Lisboa, pp. 316–331 (2021)
7. do Rosário Palma Ramalho, M.: Os limites do poder disciplinar laboral. In: Estudos de Direito do Trabalho, vol. I, Almedina, p. 188, footnote 9 (2003)
8. Geraldes, A.L.: Decision of STJ de 18.05.2017. Proceedings 4305/15.8T8SNT.L1.S1. www.dgsi.pt
9. Santos, P.: Decision of TRL de 15.12.2016. Proceedings 4305/15.8T8SNT.L1–4. www.dgsi.t
10. Mousinho, T.S.: Cliente ou fornecedor mistério e o direito do empregador a conhecer da avaliação do seu serviço e atividade professional. Prontuário de Direito do Trabalho, II, CEJ, Lisboa, pp. 369–388 (2020)
11. Chen, C., Al-Ansari, Y.: Trade secrets protection in the context of employment in China: the law and the business operations. Int. Technol. Manag. Rev. **3**(3), 179 (2013)
12. Chen, C., Al-Ansari, Y.: Trade secrets protection ..., Ob. Ci., p. 180 (2013)
13. D'Antona, M.: Whistleblowing in Italy: rights and protections for employees. Center for the Study of European Labour Law, no. 106, pp. 8–10 (2014)
14. Gomes, J.M.V.: Um direito de alerta cívico do trabalhador subordinado? (Ou a proteção laboral do whistleblower). Revista de Direito e de Estudos Sociais, Janeiro-Dezembro, Almedina, Coimbra, pp. 127–160 (2015)
15. Martins, D.C., Laga, M.: Whistleblowing and the case of Heinichvs Germany (ECHR, 28274/08): The Polishand Portuguese perspective, Labour Law and Social Rights in Europe, The Jurisprudence of International Courts, Selected Judgements, pp. 157–178. Gdansk University Press (2017)
16. Boot, E.R.: The Ethics of Whistleblowing, pp. 24–95. Routledge, New York (2019)
17. Brandão, N.: O whistleblowing no ordenamento jurídico português. Revista do Ministério Público **161**, 99–113 (2020)

18. According to article 8 of Law 93/2021, 20 December (transposition law of the Whistleblowing Directive, legal persons, including the State and other legal persons governed by public law, employing 50 or more employees and, irrespective of that, entities that are within the scope of the European Union acts referred to in part I. B and ii of the Annex to Directive (EU) 2019/1937 of the European Parliament and of the Council, hereafter referred to as obliged entities, have internal reporting channels
19. Article 351, no. 1, determines that - It is just cause of dismissal of the culpable behavior of the worker who, by its seriousness and consequences, makes immediate and practically impossible the subsistence of the employment relationship. In particular, the following causes of dismissal are a fair cause of dismissal: Repeated disinterest for the fulfillment, with due diligence, of obligations inherent to the exercise of the position or position to which he is assigned [no. 2, sub-paragraph d)]. When assessing just cause, the degree of damage to the interests of the employer, the nature of the relations between the parties or between the employee and his/her partners and other circumstances if relevant, no. 3
20. Martinez, P.R.: Incumprimento contratual e justa causa de despedimento, Estudos IDT, vol. II, Almedina, Coimbra, pp. 93–118 (2001)
21. de Brito, P.M.: Justa causa de despedimento com fundamento na violação dos deveres de assiduidade, zelo e diligência. Estudos IDT, vol. II, Almedina, Coimbra, pp. 119–134 (2001)
22. Mousinho, T.S.: Alguns reflexos da tecnologia deepfake no contexto laboral: Novos problemas (e novas soluções?). XIII European Regional Congress of the International Society for Labour and Social Security Law, APODIT (2020)

Economy, Intellectual Property and Taxation

Robot and Automation. Which are the Impacts on the Productivity, Jobs and Inequality of the Countries?

Jose Ignacio Lopez-Sanchez[1,2](✉) and Jose Luis Arroyo-Barriguete[2,3]

[1] Complutense University of Madrid, Madrid, Spain
jilopez@ucm.es
[2] GIPTIC-UCM, Madrid, Spain
[3] Comillas Pontifical University, Madrid, Spain
jlarroyo@icade.comillas.edu

Abstract. First, analysing the evolution of robots in the world in order to draw initial conclusions regarding the behaviour of some countries, then verify if countries with a higher density of robots per worker are countries whose jobs have a lower risk of being replaced by automation and are more competitive. Jobs requiring physical and manual skills, and basic cognitive skills will be the first to be automated; while the most demanding jobs will require social, emotional and technological skills. All the countries analysed need more new jobs than the ones they are going to lose. Also, we have observed that higher levels of automation lead to a reduction in inequality in the medium term.

1 Introduction

According Lopez-Sanchez et al. [1] the term **automation anxiety**, which is, the fear of being replaced by a robot in the workplace, while it may seem something typical of our time, is far from being a novel concept. On the contrary, it is a cyclical phenomenon, Bassett and Roberts, [2], that emerges periodically in public debate. In fact, we can go back to the early 19th century, when the Luddites, a group led mainly by textile workers, opposed the mechanization of their industry on the grounds that it threatened their jobs.

Automation is a reality. Many jobs disappear because they will be replaced by machines (ICT, Robots, Artificial Intelligence, among others). Companies must be prepared to face this challenge and remain competitive. If SMEs are not able to adapt to the new competitive environment they may disappear. For this reason, it is necessary that they incorporate this technology as a source of external knowledge.

The present contribution is intended to answer the next three questions:

1. Are countries with a high density of robots per worker more competitive? Is there a relationship between the risk of job automation and worker productivity, and what is the relationship with unemployment?

This study was supported by the H2020 Topic ICT 28-2017, Inclusive Robotics for a better Society (INBOTs CSA EU project 780073).

M. A. Grau Ruiz (Ed.): INBOTS 2021, BIOSYSROB 30, pp. 73–79, 2022.
https://doi.org/10.1007/978-3-031-04305-5_13

2. Will more jobs be created or destroyed in 2030?
3. Does automation increase or reduce inequality in countries?

2 Robot Density, Productivity and Unemployment

Automation itself is not bad. In fact, countries with a higher density of robots per worker are countries whose jobs have a lower risk of being replaced by automation. Hawksworth et al. [3] show a negative correlation between the potential jobs at high risk of automation, adjusted to account for industry composition, against the density of industrial robots in the country (as shown in Fig. 1).

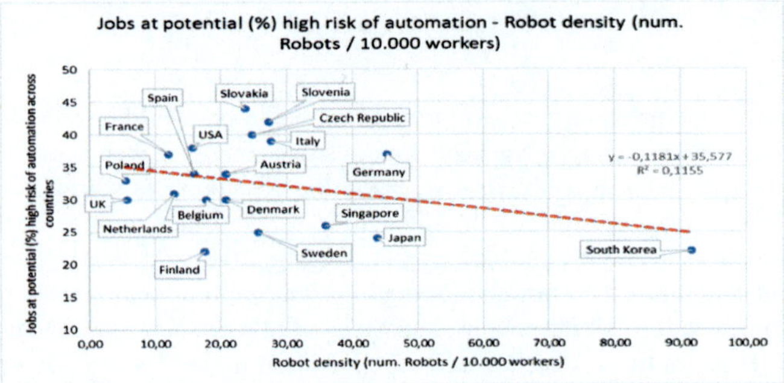

Fig. 1. Jobs at potential (%) high risk of automation across countries [3] - Robot density (num. Robots/10.000 workers).

From a first analysis Lopez-Sanchez et al. [4] do not see a clear relationship between robotization and unemployment increase. However, we can observe (Fig. 2) that those countries that have a higher density of robots, per worker, generally have a lower unemployment rate. In this case, we have considered the robot density rate, as the number of robots (all operational stock) per 10,000 workers currently working (source ILO), these data have been obtained from the International Federation on Robotics.

Similarly, in Fig. 3 and 4, those countries whose workers are more productive present a lower risk of automation of their jobs and higher robot density. As previously mentioned, countries with a higher productivity per worker are countries whose jobs have a lower risk of being replaced by automation.

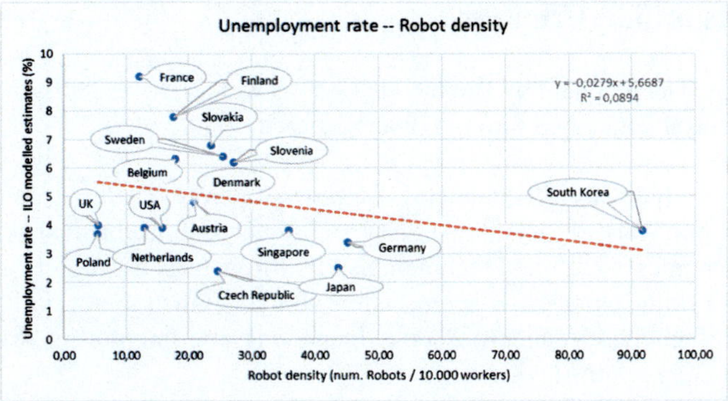

Fig. 2. Robot density (num. Robots/10.000 workers) and Unemployment rate -- ILO modelled estimates (%)

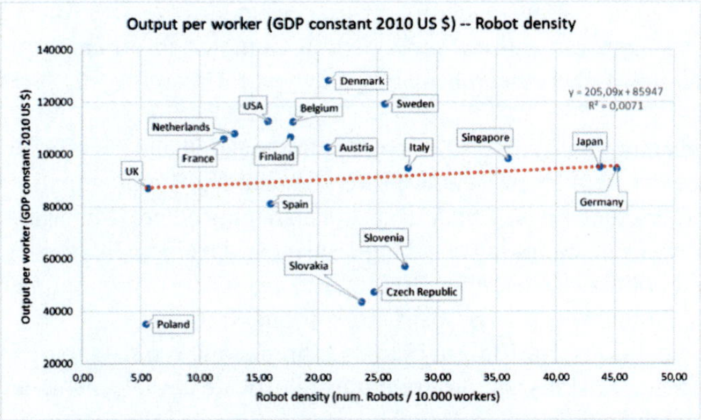

Fig. 3. Output per worker (GDP constant 2010 US $) - Robot density (num. Robots/10.000 workers).

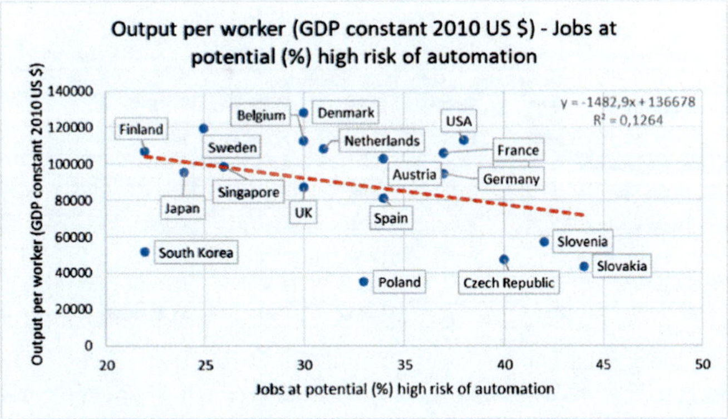

Fig. 4. Jobs at potential (%) high risk of automation across countries [3] - Output per worker (GDP constant 2010 US $).

3 Job Loss or Job Creation?

We will use the data obtained by Bughin et al. [5] as well as the 25 skills that they study and that have been contrasted through other research:

A. Physical and manual skills: General equipment operation and navigation (Drivers, assembly line workers); General equipment repair and mechanical skills (Car and truck mechanics); Craft and technician skills (Stonemasons, roofers, electricians): Fine motor skills (Nurses, food preparation workers); Gross motor skills and strength (Machine feeders, cleaners, packers); Inspecting and monitoring skills (Security guards, quality control)
B. Basic cognitive skills: Basic literacy, numeracy, and communication (Cashiers), Basic data input and processing (Typists, data entry, accounting clerks)
C. Higher cognitive skills: Advanced literacy and writing (Editors, paralegals, writers), Quantitative and statistical skills (Financial analysts, accountants); Critical thinking and decision making (Doctors, insurance underwriters); Project management (Purchasing agents, front-line supervisors); Complex information processing and interpretation (Market research analysts, lawyers); Creativity (PR specialists, music composers)
D. Social and emotional skills: Advanced communication and negotiation skills (Sales representatives, real estate agents); Interpersonal skills and empathy (Counselors, social workers, therapists); Leadership and managing others (Managers, executives); Entrepreneurship and initiative-taking (Business development, strategists); Adaptability and continuous learning (Emergency responders, programmers); Teaching and training others (Teachers, instructors, trainers)
E. Technological skills: Basic digital skills (Administrative assistants, desktop publishers); Advanced IT skills and programming (Software development, network administrators); Advanced data analysis and mathematical skills (Statisticians, research analysts); Technology design, engineering, and maintenance (Engineers, robotics experts, product designers); Scientific research and development (Scientists)

Jobs lost and gained United States (Estimation2016-2030)

Source: Own elaboration from the data provided by MGI in Bughin et al (2018); International Labour Organization (2019); OECD (2019)

United States, all sectors (1.780 hours/worker)						
Skills	**Hours worked 2016 (billions)**	**Change in hours worked by 2030 (%)**	**Hours worked 2030 (billions)**	**Change in hours worked by 2030**	**Change in Worker by 2030**	
Physical and manual skills	90	-11,00	80,1	-9,90	-5.561.798	
Basic cognitive skills	53	-14,00	45,58	-7,42	-4.168.539	**-9.730.337**
Higher cognitive skills	62	9,00	67,58	5,58	3.134.831	
Social and emotional skills	52	26,00	65,52	13,52	7.595.506	
Technological skills	31	60,00	49,6	18,60	10.449.438	**21.179.775**
TOTAL	288		308,38	20,38	**11.449.438**	

Fig. 5. Jobs lost and gained United States (Estimation 2016–2030). Source: Own elaboration from the data provided by MCI in Bughin et al. (2018); International Labour Organization (2019); OECD (2019) [4].

We can see (Fig. 5) that the United States would lose 9,730,337 jobs but would need 21,179,775 new jobs (Higher cognitive skills, Social and emotional skills, and Technological skills). The net balance would be an increase of 11,449,438 jobs [4].

Western Europe (Fig. 6) would lose 18,792,969 jobs but would need 27,605,046 new jobs. The net balance would be an increase of 8,812,077 jobs. Western Europe would lose many more jobs than the United States, and would not recover them in the same proportion [4].

Jobs lost and gained Western Europe (Estimation 2016-2030)

Source: Own elaboration from the data provided by MGI in Bughin et al (2018); International Labour Organization (2019); ODCE (2019)

Western Europe, all sectors (1.522,91 hours/worker)						
Skills	Hours worked 2016 (billions)	Change in hours worked by 2030 (%)	Hours worked 2030 (billions)	Change in hours worked by 2030	Change in Worker by 2030	
Physical and manual skills	113	-16,00	94,92	-18,08	-11.872.008	
Basic cognitive skills	62	-17,00	51,46	-10,54	-6.920.961	-18.792.969
Higher cognitive skills	78	7,00	83,46	5,46	3.585.241	
Social and emotional skills	67	22,00	81,74	14,74	9.678.839	
Technological skills	42	52,00	63,84	21,84	14.340.966	27.605.046
TOTAL	362		375,42	13,42	8.812.077	

Western Europe (Countries analysed): Austria, Belgium, Denmark, Finland, France, Germany, Greece, Italy, Netherlands, Norway, Spain, Sweden, Switzerland, and the United Kingdom. Numbers may not sum due to rounding

Fig. 6. Jobs lost and gained Western Europe (Estimation 2016–2030). Source: Own elaboration from the data provided by MGI in Bughin et al. (2018); International Labour Organization (2019); ODCE (2019) [4].

From the MGI data analysed we can make a more exhaustive analysis of five European countries with the largest working population: Germany, United Kingdom, France, Italy and Spain (Fig. 7) [4].

Jobs lost and gained by Skills per Country (Estimation 2016-2030)

Source: Own elaboration from the data provided by Bughin et al (2018); International Labour Organization (2019); ODCE (2019)

Skills: (A) Physical and manual skills; (B) Basic cognitive skills; (C) Higher cognitive skills; (D) Social and emotional skills; (E) Technological skills

Country	A	B	C	D	E	TOTAL	Lost	Gained
Spain	-1.164.548	-455.624	294.295	786.182	1.212.046	672.351	-1.620.172	2.292.523
Italy	-1.214.632	-922.420	409.382	925.188	1.374.856	572.374	-2.137.052	2.709.426
Germany	-2.609.844	-1.478.912	593.906	3.505.168	2.467.422	2.477.740	-4.088.756	6.566.496
France	-983.794	-961.706	512.164	1.306.704	1.694.221	1.567.589	-1.945.500	3.513.090
UK	-704.940	-536.051	920.330	1.882.568	1.142.280	2.704.188	-1.240.991	3.945.179

Fig. 7. Jobs lost and gained (Estimation 2016–2030). Source: Own elaboration from the data provided by MGI in Bughin et al. (2018); International Labour Organization (2019); ODCE (2019) [4].

We can see (Fig. 7) the jobs lost and won, in each of the skills, for each of the countries. All the countries analysed need more new jobs than the ones they are going to lose. This allows us to conclude that automation (robots and Artificial Intelligence) will not have a negative impact on jobs. Another issue is that countries may have workers with the skills required for the new jobs.

4 Does Automation Increase or Reduce Inequality in Countries?

Lopez-Sanchez et al. [1] using a sample of 33 European countries in the period 2000–2016, analyze the relationship between economic inequality, measured by the Gini index, and the automation level, evaluated according to the number of robots per 10,000 workers. Using a panel data approach, the conclusion is that higher levels of automation lead to a reduction in inequality in the medium term. The explanation for this phenomenon can be found in the fact that automation increases wealth in the country, which can be used by governments to reduce inequality through redistributive policies. However, it does give us an indication that the relationship between inequality and the level of automation is far from obvious.

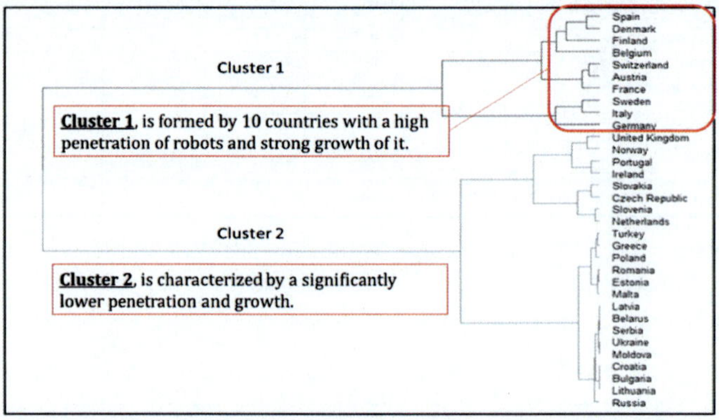

Fig. 8. Dendrogram for the time series of robot density [1]

We can identify two very different groups of contries, the first one (cluster 1), in which Germany, France, Italy, Spain are located, is formed by 10 countries with a high penetration of robots and strong growth of it (Fig. 8). The second group (cluster 2) is characterized by a significantly lower penetration and growth [1]. Countries corresponding to cluster 1 (high automation level) do not seem to be those with the most serious inequality problems. In fact, drawing the boxplot for both groups of countries (Fig. 9) shows the opposite: while inequality before taxes and transfers (gini_mkt) seems slightly higher in cluster 1 countries, once we consider inequality after taxes and transfers (gini_disp), this group of countries seems to have lower levels of income inequality. This is a first indication that, in line with the majority opinion of the IGM panel of experts, high levels of robotization generate benefits that can be used by governments precisely to reduce inequality.

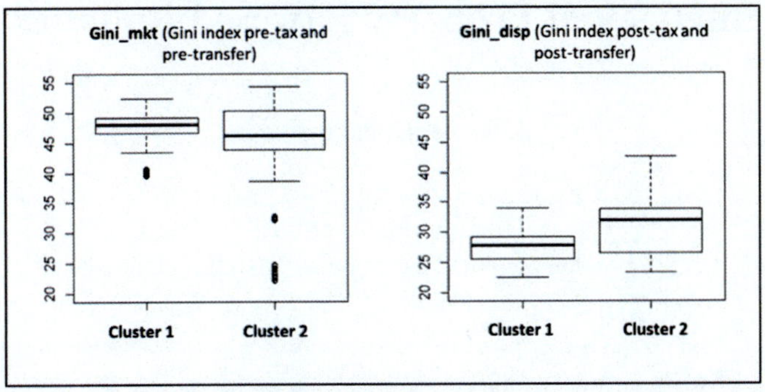

Fig. 9. Boxplot for the Gini indices, distinguishing between cluster 1 and cluster 2.J [1].

5 Conclusion

From a first analysis we do not see a clear relationship between robotization and unemployment increase. However, we can observe empirical evidence that that those countries that have a higher density of robots, per worker, generally have a lower unemployment rate. Similarly, we can observe empirical evidence that those countries whose workers are more productive present a lower risk of automation of their jobs. Countries with a higher productivity per worker are countries whose jobs have a lower risk of being replaced by automation.

We also have made a first approximation of the jobs that could be destroyed, but also that will be created. This allows us to conclude that automation (robots and Artificial Intelligence) will not have a negative impact on jobs. Another issue is that countries may have workers with the skills required for the new jobs. Also, we have observed that higher levels of automation lead to a reduction in inequality in the medium term. Automation increases wealth in the country, which can be used by governments to reduce inequality through redistributive policies.

References

1. López-Sánchez, J.I., Arroyo-Barrigüete, J.L., Curto-González, T.: Automation and inequality. UCJC Bus. Soc. Rev. **17**(4), 148–176 (2020)
2. Bassett, C., Roberts, B.: Automation now and then: automation fevers, anxieties and utopias. New Formations **98**, 9–28 (2019)
3. Hawksworth, J., Berriman, R., Goel, S.: Will robots really steal our jobs? An international analysis of the potential long term impact of automation. Price Waterhouse Coopers Report (2018)
4. Lopez-Sanchez, J.I., Grau, A., Sánchez-Urán, Y.: The impact of robotics and computerization on the labour market: inclusive insight from a Law and Economics perspective. Digit. Law Innov. Rev. **3**, 1–32 (2019)
5. Bughin, J., Hazan, E., Lund, S., Dahlström, P., Wiesinger, A., Subramaniam, A.: "Skill Shift Automation and The Future of the Workforce", McKinsey Global Institute, May 2018. https://www.mckinsey.com/featured-insights/future-of-work/skill-shift-automation-and-the-future-of-the-workforce

Intellectual Property Law and Robotics

Luke McDonagh[1,2](✉)

[1] City, University of London, London, UK
L.T.McDonagh@lse.ac.uk
[2] TT London School of Economics (LSE), London, UK

Abstract. What we may be witnessing in the not so distant future is not only the continued progress of robotic technology which will increase AI capabilities, but also an evolution of copyright and patent regimes that will put at the centre of the debate the creation of "works" or the development of "inventions", possibly diminishing the roles of "authors" and "inventors". Are we moving from the "law of authors and inventors" to the "law of copyrightable works and patentable inventions"? The robotics and AI industry may be the innovative sectors that will provide an answer to these questions.

1 Introduction

There is a growing consensus that we may be witnessing - in the not so distant future - not only the continued progress of robotic technology which will increase AI capabilities, but also an evolution of copyright and patent regimes that will put at the centre of the debate the creation of "works" or the development of "inventions", possibly diminishing the roles of "authors" and "inventors". Are we moving from the "law of authors and inventors" to the "law of copyright works and patentable inventions" regardless of who, or what, creates them? The robotics and AI industry may be the innovative sectors that will provide an answer to these questions.

2 Novel Issues

2.1 IP and Robotics

The interaction of robotics and IP rights in the context of business models – especially at SMEs – requires legal clarifications. Patents, trade secrets, copyright, trademarks and designs are tools that robotics companies can use to attract, and recoup, the big investments that are needed in this field.

While there are phases within the life cycle of robotic firms where a cooperative and non-IP focused policy is better suited to support the growth of the venture, IP strategies based on the exploitation of proprietary rights are certainly key in shaping and strengthening this industry. Just filing a patent, design or trademark application may not be enough – taking legal action against competitors that try to free-ride on the

This presentation was supported by the Horizon2020 INBOTS project.

M. A. Grau Ruiz (Ed.): INBOTS 2021, BIOSYSROB 30, pp. 80–83, 2022.
https://doi.org/10.1007/978-3-031-04305-5_14

investments made by someone else is also increasingly necessary. We have mentioned some of these lawsuits, several of which have been settled with a consent judgement that has left the complainant that enforced its IP in a stronger position than the alleged infringer. Yet, some legal actions aimed at protecting IP in robotics have also been criticised, particularly where patents have been granted by patent offices with broadly drafted claims and aggressively enforced against competitors.

2.2 Smart Robots

Robots, especially those embedded with AI, may also be capable of coming up with works or inventions that are usually created by human beings, be they painters, illustrators, poets or engineers. Whether these artistic, literary, musical or technical outputs meet the requirements for copyright or patent protection, and who the author or the inventor is, are thorny issues that are currently giving IP scholars and lawyers food for thought: issues that will become even more burning as technology keeps advancing and allows robots and machines to perform behaviours or tasks with a higher degree of autonomy.

What we may be witnessing in the not so distant future is not only the continued progress of robotic technology which will increase AI capabilities, but also an evolution of copyright and patent regimes that will put at the centre of the debate the creation of "works" or the development of "inventions", possibly diminishing the roles of "authors" and "inventors". Are we moving from the "law of authors and inventors" to the "law of copyrightable works and patentable inventions"? The robotics and AI industry may be the innovative sectors that will provide an answer to these questions.

The use of smart robots and inventive AI systems raises a range of novel issues which the existing patent framework may not be able to accommodate easily. It is unclear whether AI algorithms should be subject to patent protection, or whether the proliferation of AI-generated inventions should require us to re-evaluate core patent concepts such as inventive step. There is no consensus over whether a machine can be designated as inventor. Who would be the owner of the patent where an AI system has generated the invention with little or no human intervention? Who would be liable for AI-induced infringement in cases where machines act "autonomously"?

3 The Field of Robotics

The field of robotics is remarkably wide, with many social settings now entailing and increasingly requiring the use of robots to support a variety of human activities. Unsurprisingly, robots' form and shape, their level of intelligence and intended purpose can vary significantly depending on the relevant industry.

Domestic robots are already a reality in a growing number of family homes. They include both humanoid robots which support those in need (such as the elderly, people with disabilities or children) and robots for household consumer markets, including domestic vacuum cleaners and grass- trimmers.

Humanoid robots only account for a small fraction of the industry with robotic arms for industrial automation being instead widespread. Furthermore, innovation in artificial intelligence (AI), processing power and sensors have all helped to enhance the field of robotics and increased the type of tasks that robots can perform. Robotics companies can use AI to make their final products more functional, and future improvements and incorporation of AI into robots will make them more interactive, accepted and appreciated.

First, robotics R&D and innovation are highly concentrated in a handful of countries, such as the US, EU (especially Germany) and Japan. Available patent data also reflects this geographical concentration, with applicants from the above countries accounting for the vast majority of patent applications in the field. Second, the ecosystem consists of a network of different actors, e.g., individuals, universities and research institutions (with their spin-out companies) as well as technology-intensive firms, all collaborating to drive innovation forward. Third, government funding is often key to supporting robotics R&D.

IP protection is vital in all R&D intensive industries, and the field of robotics is no exception. Robotics firms often undertake years of intensive (and expensive) research before being in a position to sell their products and reach commercial success.14 The lengthy and costly process of delivering profitable products highlights the role of IP rights which are viewed as necessary to recoup up-front investments and to fend off competitors seeking to capitalise on the R&D investments of their rivals.

Any company subject to due diligence because of, for instance, a strategic investment plan, acquisition or IPO, will likely have its IP portfolio reviewed as part of this process, with potential investors likely to view robotics firms without a strong IP portfolio as a less attractive investment option. While there are phases within the life of robotics firms where a cooperative and non-IP focused policy is better suited to support the growth of the venture, IP strategies based on the exploitation of proprietary rights are certainly key in shaping and strengthening this industry. Just filing a patent, design or trademark application may not be enough – taking legal action against competitors that try to free-ride on the investments made by someone else is also increasingly necessary.

Patents, trade secrets, copyright, trademarks and designs are tools that robotics companies can use to attract, and recoup, the large investments that are needed in this field. The patent route can be particularly valuable for companies whose robots, or their elements, can be easily reverse-engineered (reverse-engineering is the process whereby a product can be deconstructed to disclose its elements and the way it is manufactured). Indeed, in situations where reverse-engineering is straightforward, filing for a patent may be favoured over the alternative tactic - trying to protect the process of manufacturing and/or the relevant product by keeping them secret - with that patent being enforceable against any third party that exploits the invention without the patentee's consent. Symmetrically, relying on trade secrets to protect robotics inventions can work well where: (i) robots are produced and used in a controlled environment; (ii) reverse-engineering is not easy to carry out; and (iii) those working with the products are committed to secrecy. Trade secrets (rather obviously) do not require disclosure, as the patent system does. A patent (and the monopoly coming with it) is granted in return for the disclosure of technical information so that the public at large, including patentees' competitors, will be able to exploit the invention after the 20-years term of protection expires. Certain elements of robotics devices, especially software code, could be protected by copyright if they satisfy the relevant requirements, including originality Software is crucial in this field, with robots being unable to function without underlying code – robots deprived of software would basically be unable to perform their intended tasks. What about trademarks? How can this IP right add value to robotics companies and their products? In general, registering trademarks is crucial to protect products' goodwill and reputation, especially in business-to-consumer industries. Notably, robotics is increasingly becoming an industry where products are sold directly to millions of end-users (consumers). The commercial success of products such as nanny-robots, pet-robots, caretaker-robots and medical-robots also depends on a reliable brand which consumers know, trust, appreciate and remember. Meanwhile, the exclusive rights given by design registration can be enforced against third parties that use designs that are perceived by an informed user as giving the same overall impression.

Within the overall IP field, the decision to apply for a patent, rather than simply utilise these other IP rights, may be influenced by the complexity of the company's products and whether the company's competitors are likely to get their hands on such products and subsequently reverse-engineer them.

While there are phases within the life cycle of robotic firms where a cooperative and non-IP focused policy is better suited to support the growth of the venture, Are we moving from the "law of authors and inventors" to the "law of copyright works and patentable inventions"? This raises a follow on question. Should these outputs be protected by copyright and patent laws? Should they be left to the public domain? If they are protectable, who should be deemed the owner of the resulting copyright or patent? The programmer? Or the user? The robotics and AI industry may be the innovative sectors that will force law makers and judges to provide answers to these questions.

4 Conclusion

In conclusion, it is recommended that a number of new legal reforms, should be enacted at the European level. These should be brought forward to bring a measure of clarity to this rapidly developing area. In the areas of copyright and design law, where the EU has substantial competence, new legislative guidance on (i) whether interactive robot-created works of e.g. music, literature, art can be protected by copyright and (ii) if so, who should own these works? Similar reforms could be brought forward by the European Patent Office to clarify the circumstances regarding (i) whether an interactive robot-created invention is patentable or whether a human inventor is required; and (ii) who should be the owner of such a patent (e.g. the consumer/user of the robot, the developer of the robot)?

Check for updates

Robot-A(I)rtists and Tax Provisions

L. Scarcella(✉)

DigiTax Centre, University of Antwerp, Antwerp, Belgium
luisa.scarcella@uantwerpen.be

Abstract. The use of new technologies to produce art is part of the physiological evolution of art itself. Like the camera contributed to shaping the course of art, it is fascinating to witness how new technologies will mark the future of art. However, using deep learning and other AI techniques to create art leads to many new questions: who is the real author of the work? How should art or creativity be attributed? Can AI make art in its own right? And if yes, can these art pieces fall under the definition of work of art as included in already existing tax provisions? While it is difficult to predict how the deployment of AI will evolve in the following years, existing prototypes give us food for thought on how the current legal framework could be challenged. This chapter offers a primer of the tax questions that emerge in this field.

1 Introduction

In summer 2021, the Design Museum in London hosted an exhibition titled "Ai-Da: Portrait of the Robot", showcasing Ai-Da's work. Named after mathematician and computer pioneer Ada Lovelace, Ai-Da is not the artistic nickname of a visionary futuristic artist but the name of an AI robot that can create artistic pieces from sight [1]. Ai-Da draws and paints using cameras in her eyes, AI algorithms and her robotic arm. Artificial neural networks create the foundation for Ai-Da's abstract paintings and subsequently, (human) artist Suzie Emery applies oil paint to canvas giving life to the final works [2].

Three years earlier, in 2018, Christie's auction house in New York became the first auction house to put an artwork generated by an algorithm under the hammer. They auctioned an artwork created by Paris-based art collective Obvious using artificial intelligence. The piece - a portrait generated by a machine learning (ML) algorithm called a generative adversarial network (or GAN) [3] - has been sold for nearly half a million dollars. More than 45 times the estimated selling price of $7,000 to $10,000[1].

These are only two examples of how AI and robotics are changing the world of art, but artists have long integrated new technologies into their practices, with AI being no exception. Nevertheless, there is a crucial difference. With new evolvements in AI and machine learning, we are starting to question whether the machine can be its own source of creativity, creating bewildering and unique novel art pieces thanks to the combination of vast amounts of historical and social data [4]. To what extent is a machine a tool, a partner or a substitute for artists? Like cameras, AI can offer new means for artists

[1] https://www.christies.com/features/a-collaboration-between-two-artists-one-human-one-a-machine-9332-1.aspx.

M. A. Grau Ruiz (Ed.): INBOTS 2021, BIOSYSROB 30, pp. 84–92, 2022.
https://doi.org/10.1007/978-3-031-04305-5_15

and non-artists alike to express themselves artistically. But can an artificially intelligent machine be an artist in its own right? And can its outcomes be considered works of art under existing tax provisions?

2 Robots, AI and the Arts

Recent developments have led to machines that can independently learn and create, making the human input in the creative process more redundant. Ai-Da is just one of the many developments of AI with the intent of enabling it to create its art. Previously, scholars from other legal disciplines involved in the INBOTS project have referred to robot-artists such as the "Painting Fool" or "Paul" [5]. "The Painting Fool" is a computer program which can simulate the physical painting process. By detecting people's emotions, it can paint portraits and invent visual scenes by means of generative techniques [6]. Another example is the above-mentioned Paul, an interactive robotic installation which can create people's portraits by using its camera eye. Differently to the Painting Fool, which is a computer program, Paul consists of a physical robotic arm [7]. At the same time, AI and robotics are not only deployed in the creativity realms of fine arts but their usage in the artistic domain transcends all forms of artistic expressions. Recently, the use of AI has been key in completing the unfinished Tenth Symphony by Beethoven. All that remains of Beethoven's Tenth Symphony were only fragmentary sketches of the first movement which he started before his death in 1827. Thanks to the use of AI, a group made up of music historians, musicologists, composers and computer scientists were able to turn sketches into a complete piece of music addressing the challenge of ensuring that the final outcome would remain faithful to Beethoven's process and vision [8]. However, research, experiments and prototypes have been developed looking at how AI could produce its own novel music pieces [9, 10] or write stories [11].

The trend of AI and robotics deployed in the art domain is only increasing. However, as it emerges from all these different examples, the level of involvement and autonomy of algorithms and machines can be very different. Indeed, when looking at art made through or by robots and AI, a distinction should be made between computer-aided and computer-generated works [5]. The first category includes works that are produced by humans with the mere help of machines. The second one entails a higher degree of autonomy by the machine referring to output autonomously created by AI. However, drawing the crossing line between what is computer-aided and computer-generated can be very difficult to individuate. In the case of Paul, it seems easier to sort out our doubts since Paul's robotic arm represents a complement to the human artists. More difficult is the case of AI-Da or of the "Painting Fool", striking examples of higher level of machine autonomy.

3 Reasons Why Robots and AI Creativity Becomes Relevant in the Tax Realm

3.1 The Concept of Work of Art in VAT Provisions

From a tax viewpoint, the concept of art becomes relevant in different tax provisions and types of taxes. Within the European context, the European VAT Directive contains specific provisions applicable to works of art. On the one hand, some provisions target intra-community exchanges and aim to avoid double taxation. On the other hand, other provisions refer to the possible application of reduced rates to works of art based on a value-based consideration of societal importance of art.

Against this backdrop, Article 103 of the European VAT Directive states that the Member States may decide to apply the reduced rate to the supply of works of art in two cases. In the first case, these art pieces are directly supplied by their creator or successors in title. In the second, when, on an occasional basis, the supply of the works of art is by a taxable person other than a taxable dealer and where the works of art have been imported by the taxable person himself, or where they have been supplied to him by their creator or his successors in title.

Many EU Member States have opted to apply a reduced rate to the transactions involving these objects. For instance, this is the case of Italy which applies a 10% reduced rate or France, where the VAT rate drops to 5.5%. Nonetheless, the boundaries of the scope of application of the VAT reduced rate are set according to what shall be understood as a work of art.

The VAT Directive does not offer a general definition of "work of art"; instead, it includes a list of objects to be considered as such contained in Annex IX of the VAT Directive. This list contains objects such as paintings and drawings (including drawings for architectural, engineering and industrial purposes), hand-decorated manufactured articles, engravings, sculptures, tapestries, pieces of ceramics. However, the objects mentioned above are all accompanied by the requirement that they need to be executed entirely by the artists. Furthermore, in the case of points (1) (2) (4) and (6), which include paintings, tapestries and enamels on coppers, the requirement is even stricter as these objects need to be made exclusively by hand by the artist. The Annex also often specifies the limited number of production or artist's signature requirement for each of these art pieces. As the objects in question are classified as 'works of art' for the purposes of the application of the VAT system, their creators are classified as 'artists' consequently. Hence, according to the VAT Directive, works of art are then defined only by their inclusion on the list. It derives that it is the inclusion of the objects in the list that confers the title of artist to its creator and not vice-versa.

3.2 The Concept of Work of Art in Direct Tax Provisions

Concerning direct tax provisions, works of art have been the targets of different beneficial tax treatments. Some States allow the payment of due taxes through the handing over of works of art, such in the case of inheritance taxes or income tax. At the same time, tax reductions can be granted for donations of works of art to public art institutions. One of the main goals that has been pursued through the adoption of these types of policies is the creation of a mechanism through which objects of exceptional cultural value enter public collections to be enjoyed by thousands across the country. However, this mechanism represents a win-win solution. On the one hand, the State can receive and protect the artistic heritage, which until then has been preserved in the homes of private collectors. On the other hand, families would avoid losing liquidity by handing art pieces to pay due taxes.

France was one of the first countries in the European Union to introduce provisions allowing the payment of inheritance taxes with works of art in 1968. Designed by André Malraux in 1969 when he was Minister of Culture, the law n°68-1251 of December 31, 1968 and its implementing decree n°70-1046 of November 10, 1970 introduced the "dation en paiement", which has considerably enriched the State's collections [12]. This law allowed the payment of the inheritance tax debts through the handing over of works of art, books, collectables, and documents of high artistic or historical value. The "dation" can only be proposed after the taxable event has occurred by the heirs and legates, and cannot be used to pay in advance future tax debts that are not certain in principle or whose amount is still to be set. Moreover, the procedure is only applicable when the amount of the duties that the interested party proposes to pay by "dation" is at least equal to €10,000 for each due tax. The value of the property proposed for donation to the State may represent all or only part of the duties owed by the applicant, and the "dation" can only concern goods of which the taxpayers are full owners. Moreover, the offer of dation is not admissible for property held for less than five years by the interested party unless the owner entered in possession of the art piece by gratuitous transfer. Finally, the offer of dation is not admissible if the goods offered in payment have previously given rise to 2 refusals of approval.

From a procedural viewpoint, this mechanism requires the approval of the Minister of Economy and Finance on the proposal of the competent minister and the advice of a commission. The approval sets the property's value and must be accepted by the heir. It is possible to set up an expertise procedure in case of disagreement with the State on the value of the work. Hence, assessing what constitutes an asset of high artistic or historical value becomes essential. Based on the work of the interministerial commission in charge of examining the offers of art pieces for the payment of due taxes, it is either about works realized by preeminent artists as L'Astronome by Vermeer, works of art presenting a sociological, historical, political or religious testimony or art pieces by rediscovered artists.

The scope of this tax payment mechanism was subsequently extended in 1982 to the solidarity tax on wealth (ISF) and now applies to the tax on real estate wealth (IFI). The "dation en paiement" has enabled works of exceptional quality to enter the national heritage, including Vermeer's Astronomer, Fragonard's Portrait by Diderot, which have enriched the collections of the Louvre, while the Caillebottfe's Landscape at Argenteuil, Courbet's Origin of the World, Claude Monet's Luncheon on the Herb, Degas' Dancers were acquired and are still at display at the Musée d'Orsay. Other works of art that are now property the State are: André Breton's Wall of Objects, Henri Matisse's Nude with Oranges, which are currently housed at the Musée National d'Art Moderne as well as major works by Bacon, Braque, Calder, and Rothko. The list is long and it includes also the treasure of Rethel (a masterly collection of Gallo-Roman gold and silver), the treasure of Roman gold and silverware from Boscoreale, royal furniture, a large part of the archives of Viollet-le-Duc and Claude Lévi-Strauss, but also manuscripts by Montesquieu, Jules Verne, Marcel Proust, Simone de Beauvoir and Jean-Paul Sartre. The correspondence of Debussy as well as musical instruments have also all been acquired by the State through this mechanism [13].

Similarly, following the example of France, in Italy, articles 6 and 7 of Law no. 512 of 1982 (the so-called Guttuso Law named after the painter Renato Guttuso) introduced the possibility of paying income and inheritance taxes through the transfer of works of art. In order to make the payment through the transfer of works of art - as highlighted by the Financial Administration with Resolution no. 347 of August 5, 2008 - it is necessary that the tax debt is a liquid debt, certain and collectable, since similarly to what is foreseen in France, this mechanism does not apply to possible future payment obligations [14]. The law initially provided a lengthy procedure for both types of taxes (income and inheritance taxes). In order to start this procedure the taxpayer (or their heirs and legatees) will have to submit a proposal accompanied by a detailed description of the asset offered, the value deemed appropriate and the ownership of the same. The procedure involves the expertise and approval of the Ministry of Culture and the Ministry of Finance in different stages. However, this procedure resulted in a lengthy bureaucratic process leading to a limited use of this mechanism. Official data dating back to 1991 show that among the 50 proposals received by the Ministry of Culture, only 14 had a positive outcome. Hence, the procedure was renovated in 2014, with the creation of an interministerial commission. This change recognized the potential of this mechanism, and thanks to this renovated procedure, in 2017, the Ministry of Cultural Heritage and Activities purchased the Richard Ginori Museum of the Doccia Factory in Sesto Fiorentino, one of the most important ceramic and porcelain manufacturers [15].

In the case of Belgium, both the Inheritance tax code of the Brussels and Walloon Region (art. 83/3) and the Flemish Tax Code (art. 3.4.3.0.2) foresee that inheritance taxes can be paid by handing over works of art if these art pieces have an international belonging to the cultural heritage or are international renown. Moreover, article 145/33 of the Belgian Income Tax Code grants a tax reduction for donations to institutions recognized by the Federal authorities, including works of art [16].

In other countries, such as Switzerland, the "la dation en paiement" was only adopted at the cantonal level by the cantons of Jura, Geneva and more recently Vaud. The procedure is very similar to the French one. The taxpayer proposes to pay the tax by handing over cultural properties of "high artistic, historical or scientific value" (Geneva and Jura) or a "major importance for the canton" (Vaud) [17]. A recent example of acquisition of works of art by public museums through the use of this mechanism can be found in the case of Vaud canton. In 2017, the collections of the Musée Cantonal Vaudois des Beaux-Arts in Lausanne were enriched with seven drawing books by Aloïse Corbaz, one of the most influential female artists of the outsider art/art brut [18].

Outside Europe, an important example is South Korea, where Samsung Electronics reached an agreement with the government to pay its due taxes by transferring several works of art. In this case, the South Korean government acquired an immense artistic patrimony by receiving over 23 thousand works of art for an estimated value of around 9 billion dollars. Among the works ceded to the State, there are 14 pieces included in the list of national treasures of South Korea, works by Marc Chagall, Pablo Picasso, Paul Gauguin, Claude Monet, Joan Miro and Salvador Dali [19].

The different policies adopted by the countries mentioned above have a common denominator: the level of discretion of the government when determining what constitutes a work of art of high cultural or historical or international renown value. This level of discretion at governments' disposal opens up questions on how the commissions and other organs in charge of evaluating the cultural and artistic value of art pieces will interpret the work created by or with the assistance of robotics and AI.

3.3 Can Robotic/AI Art Fall Under the Tax Concept of Work of Art?

As AI and robotics are starting to take the stage in the art world, questions might arise on the applicability of tax provisions which refers to the concept of work of art as reported in the previous sections.

In relation to the concept of work of art as included in the EU VAT Directive, the list of objects contained in Annex IX often refers to the requirement that the artefact is executed entirely by the artist or even entirely by hand by the artist (e.g. for paintings and drawings). While at European level, there is no decision on whether the use of new technologies by an artist undermines the possibility of applying reduced VAT rates, some tax authorities have already issued some guidance on the matter.

In 2020, the Italian tax authority provided a public answer to a ruling concerning the application of the VAT reduced rate on sculptures created through 3D printers [18]. More specifically, concerning sculptures, the Annex IX contained in the VAT Directive states that these are to be considered as works of art only provided that original works of statuary or sculpture, in any material, are executed entirely by the artist and if the casting of sculptures is in the limited edition amounting to eight copies. According to the Italian Tax authority, the conditions required for the application of the preferential VAT rate do not apply when these are not performed entirely by the artist but are made in whole or in part through the use of mechanical processes, such as 3D printers, FDM 3D modelling software, as well as with the intervention, in some cases, of other parties who make use of the same technology [18]. Consequently, following the same reasoning of this ruling, it can be expected that a similar approach would be taken in the case of works of art which involve in their realization process the use of AI and robotics when Annex IX requires that these objects are realized entirely by the artist.

Nonetheless, this understanding of art production could quickly become obsolete as more and more artists explore the use of new technologies in their art as representative of our society and the times we live. The use of new technologies in the art world is physiological. Concerning photography, the Advocate General, in its opinion on the *Photographique* ruling, has already stated how focusing on the technical differences (e.g. analogue technology versus digital technology) that do not affect the artistic character of the photography might pose severe risks of distortion of competition [19]. Similarly, we can wonder how creating a discriminant between the techniques used, and the involvement of technologies can ultimately lead to forms of discrimination among artists in the world of fine arts.

Differently, admitting that a machine entirely creates a work of art without any input by humans will make the machine the artist. Nonetheless, the reduced rate applies to the supplies by their creator. Given that the CJEU has already clarified that the words creator and artist are synonyms in relation to the concept of "work of art" within the VAT Directive [20], it is hard to imagine how – within the given legal framework – a robot or AI could be considered as a taxable person.

As it concerns the application of direct tax provisions allowing the payment of due taxes through the transfer of works of art, the analysis of what constitutes a work of art and whether the art created by the robot or AI – autonomously or complementarily – depends on a case-by-case analysis based on the requirements set by the national legal framework.

Often it is required that the work of art is of great artistic or historical value for the cultural heritage of that particular country or region. Sometimes this evaluation is delegated to a ministerial commission including art specialists, which will offer a qualitative assessment of the work of art. This means that the discretion level that already characterizes the evaluations of works of art produced by human artists can be transposed to the art pieces made by robots or artificial intelligence. Thus, depending on a continuously evolving sensibility towards what society understands as art.

Consideration in the area of intellectual property could be relevant if the commissions in charge of the evaluation of art pieces and other stakeholders such as courts, tax administrations and taxpayers should look for guidance in other areas of the law to determine what constitutes a work of art. Indeed, the underlying element of all tax questions resonates with what has been sustained in the area of copyright, namely that the key IP issues are essentially the same whether we talk about pure and intangible AI-empowered software (e.g. algorithms) or if we consider AI-equipped interactive robots that rely on physical embodiment to enhance creativity. IP scholars have already started to address whether outputs from machines and AI can be protected by copyright and patent laws and, in case of a positive answer, who should be deemed the owner of the resulting copyright or patent, the programmer or the user? [5].

In order to be protected by copyright laws, there are two main elements to be considered: originality and authorship. Regarding the originality element, according to what was previously established by the CJEU, "works such as computer programs, databases or photographs are protected by copyright only if they are original in the sense that they are their author's own intellectual creation" [21]. In a later decision, the court stated that an intellectual creation is an author's own if it reflects the author's personality by expressing its abilities in the production of the work by making free and creative choices which would enable them to stamp the work with their "personal touch" [22]. Emphasizing the author's "own intellectual creation" and "personal touch" seems to imply the necessity of a certain degree of human authorship to fulfil the originality requirement [5]. This interpretation of the originality requirement becomes relevant when looking at the distinction between computer-aided and computer-generated works. As noted by the Advocate General in the *Eva-Maria Painer* decision, human creations are protected even when the person employs a technical aid [23]. Thus, computer-aided works would fall under copyright protection. More complicated is the case of computer-generated works, which would not qualify as "human creations". However, in the above-cited decision, the CJEU recognized that an author could stamp a portrait photograph with her personal touch by making choices such as the applicable background, lightning, angel and atmosphere of the portrait. Scholars have already argued how it can be doubted that these machines do not make creative choices. For instance, in the case of the painting fool, the machine would choose adjectives to select a filter to achieve an appropriate visualization. The program also selected one of seven rendering styles involving the simulation of paints, pencils and pastels to produce the portrait [3].

Moreover, there is heterogeneity among the different copyright laws concerning the authorship elements, which often limit authorship to natural persons. According to Spanish law, an author is a natural person creating the work. France requires a natural person to be the author as well, and the US Copyright Office also emphasizes the importance of the human element to register original work of authorship. The human element does not raise difficult questions in relation to machines which only complement the creative process. However, in the case where the machine outputs are prevalent to the human ones leads to the need for a case-by-case analysis to determine the predominance in the creative process considering whether the work will predominantly reflect the human author's own intellectual creation. If from this type of analysis, the machine outputs outweigh the human ones, the human element will be considered as lacking, and copyright will not subsist.

Even if substantial research on the topic has been made in the area of IP law, doubts still persist and can be reflected in the tax realm as well.

4 Conclusion

The definition of creativity has intrigued philosophers since the dawn of time. Understanding what art is and how it sparks has been a centuries-long restless quest. At a time when governments all around the world have recognized the societal value and importance of promoting the arts - also through the adoption of tax provisions in this direction - new developments in AI and robotics bring one more time to the fore the pursue of trying to conceptualize creativity. Determining whether AI can realize or not proper work of art and its complementary contribution to the realization of art by human artists open up new challenges also in the legal field. While IP scholars have been studying in depth the subject for already some years, the applicability of existing tax provisions has so far been underestimated. Nonetheless, in our current tax frameworks, there are several provisions that refer to the concept of art in indirect, as well as in direct tax matters. Questions on the applicability of these provisions arise both in the context of computer-aided art pieces and in the more forward-looking autonomously computer-created art. A case-by-case analysis of the type of technology deployed and its level of involvement in the creation of works of art is the first step to determine which provisions might be applicable and to understand possible limits of our legal and tax framework.

References

1. Romic, B.: Negotiating anthropomorphism in the Ai-Da robot. Int. J. Soc. Robot. (2021). https://link.springer.com/content/pdf/10.1007/s12369-021-00813-6.pdf
2. Ambrosio, C.: Unsettling robots and the future of art. Science **365**(6448), 38–39 (2019)
3. Goodfellow, I., et al.: Generative adversarial nets. In: Advances in Neural Information Processing Systems, vol. 27, pp. 2672–2680 (2014)
4. "Clarifying the role of IP law", chap. 3, Inbots White Paper, Interactive Robotics. Legal, Ethical &. Socioeconomic Aspects (2019)

5. Colton, S.: The painting fool: stories from building an automated painter. In: McCormack, J., d'Inverno, M. (eds.) Computers and Creativity, pp. 3–38. Springer, Heidelberg (2012). https://doi.org/10.1007/978-3-642-31727-9_1
6. Tresset, P.A., Fol Leymarie, F.: Portrait drawing by Paul the robot. Comput. Graph. **37**(5), 348–363 (2013)
7. Muñoz Lago, P.: A Deep Learning Approach to generate Beethoven's 10th Symphony" (2019, unpublished)
8. de Mantaras, R.L., Arcos, J.L.: AI and music: from composition to expressive performance. AI Mag. **23**(3), 43 (2002)
9. Zulić, H.: How AI can change/improve/influence music composition, performance and education: three case studies. INSAM J. Contemp. Music Art Technol. **1**, 100–114 (2019)
10. Smith, L.: Will AI write the next Great American Novel?" California Magazine, 15 April 2021
11. Bouglé, F.: Art et transmission patrimoniale [deuxième partie]. Revue Juridique de l'Ouest Année **2**, 131–173 (1999)
12. Perrotin, F.: La dation en paiement, un dispositif encore méconnu. LPA, issue 152t2, n° 152t2, p. 3, 3 June 2020
13. Quattrocchi, A.: Il Pagamento Delle Imposte Dirette Mediante Cessione Di Beni Culturali. Diritto e Pratica Tributaria, vol. 1, p. 134 (2020)
14. Il ministero dei Beni Culturali acquista (e salva) il Museo Richard Ginori, Il Sole 24 Ore, 27 Novembre 2017
15. Couturier, J.J., Peeters, B., et al.: Belgisch belastingrecht in hoofdlijnen, 26 edn, p. 9 (2020)
16. Renold, M.A.J.: Le droit de l'art et des biens culturels en Suisse: questions choisies. Revue de droit suisse, vol. 129. Halbbd. 2, n° 1, pp. 137–220 (2010)
17. Musée cantonal des Beaux-Arts: "Le mcb-a enrichit son fonds Aloïse avec 8 nouveaux cahiers", 17 January 2017. http://www.emailing.vd.ch/wb.php?p=1nj/1ef/rs/1jd/tp/rs
18. Response to interpellation No. 303 of 2 September 2020
19. Case C-145/18 Regards Photographiques SARL v Ministre de l'Action et des Comptes publics, Opinion of the Advocate General, 7 March 2019, at 39
20. Case C-145/18 Regards Photographiques SARL v Ministre de l'Action et des Comptes publics, 5 September 2019
21. Case C-5/08 Infopaq International A/S v Danske Dagblades Forening at 35
22. Case C-145/10 Eva-Maria Painer v Standard VerlagsGmbH and Others at 92
23. Case C-145/10 Eva-Maria Painer v Standard VerlagsGmbH, Opinion of Advocate General Trsteniak, 12 April 2011, at 121

Taxing Robots

Rita de la Feria[1(✉)] and María Amparo Grau Ruiz[2,3(✉)]

[1] Tax Law, University of Leeds, Leeds, UK
R.DeLaFeria@leeds.ac.uk
[2] Financial and Tax Law, Complutense University of Madrid, Madrid, Spain
grauruiz@ucm.es
[3] Transnational Taxation, Northwestern University, Evanston, IL, USA

Abstract. In recent years, the idea of taxing robots has been progressively gaining momentum. The potential impact of automation on employment, and consequently on income tax revenues, has led many to defend the introduction of a tax on robots, or on the use of robots, to either compensate for the potential revenue loss, or to slow down the process of automation. This paper argues that whilst automation presents significant challenges to tax systems, the introduction of a new tax on robots –or on their use– are not effective mechanisms through which to address these challenges. In order to explain the growing popularity of *taxing robots*, the paper draws insights from behavioral science. It concludes that the growing support for such a tax is more a reflection of unconscious and institutional biases, than it is of sound taxation principles.

1 Introduction

We met Boris in March 2017, at a tax law conference in Barcelona. He was extremely knowledgeable, and responded with confidence and precision to all the audience's questions on Spanish tax law; we had no doubt that he had a bright future ahead. Except, Boris was a smooth, 1.5 m high, anthropomorphised, robot. Our experience mimics that our others in legal practice [1] – robots are coming.

There is now strong evidence that ongoing developments in robotics will have a significant impact on the labour market [2], with a growing literature on the nature, magnitude, and consequences of those changes. There is a growing consensus on the long-term effects of robotics and automation, which are estimated to include increased labour demands through higher productivity (albeit experienced asymmetrically) [3], as well as addressing demographic challenges, such as ageing populations [4]. In the short-term, however, significant challenges are likely to arise. In particular, there is a displacement effect that may outweigh these positive impacts, leading to an overall decline in employment and wages [5, 6]. These displacement effects are most evident

This paper was presented at conferences, or seminars, held at Northwestern University, the Confederation of British Industry (CBI), London, and Lund University. We are grateful to the organisers and participants at those events for the comments received therein. The authors also gratefully acknowledge funding from Horizon 2020, *INBOTS: Inclusive Robotics for a Better Society*, project 780073.

M. A. Grau Ruiz (Ed.): INBOTS 2021, BIOSYSROB 30, pp. 93–99, 2022.
https://doi.org/10.1007/978-3-031-04305-5_16

on specific industries, such as manufacturing [3], specific workers, namely low-skills workers [7], and specific regions [8]. Further, it is also likely to increase over time, due to the "superstar effect", according to which output is relocated to firms with lower and declining labour share [9]. Indeed, even for those who remain in work, their conditions are also likely to change, with workers expected to see a decrease in both wages and in working hours resulting in depressed incomes despite the increase in higher-paid jobs [10]. Certainly, automation is not a new phenomenon, and neither are societal anxieties around technological advance and change [11, 12].

Nevertheless, the ongoing developments in the labour market are regarded as a cause for concern, not only because the speed of changes raises fears over the capacity of the world economies to adapt, but also because great leaps in automation have historically given rise to significant [economic?] upheaval [13].

From a public finance perspective, these fundamental labour changes are likely to be hard hitting, through a double- effect. On the one hand, labour displacement is likely to have a negative impact on revenues from personal income taxes, which are unlikely to be compensated by any potential increase on corporate income taxes. On the other hand –and occurring at the same time– both labour displacement and replacement are likely to require increased expenditure commitments. This will include unemployment benefits and the need to re-skill the work-force, but also the possible creation of an unconditional basic income (UBI) [14].

Acknowledgement of these challenges has led many, including Bill Gates in a now famous interview, to defend the introduction of a tax on robots or their use [15]. In 2017, the proposals gained a new momentum, with a Report from European Parliament, openly considering the possibility of either a robot tax, or a tax for using robots [16]. Since then, proposals for taxation of robots, or of their use, have been progressively gaining traction. Under these proposals, the new tax would have either a proxy or a regulatory aim – or both. A proxy tax, to the extent that it would compensate for the potential revenue loss in personal income taxes to create an equal playing field between humans and robots [17, 18], and a regulatory tax to internalise the negative externalities of robot use and slow down the process of automation [19].

In what follows, we work to bring critical nuance to the debate around taxing robots. In Section II, we consider the tax rationale for these proposals and their significant limitations, arguing that the introduction of a new tax on robots –or their use– is not an effective method of addressing the challenges that automation presents for tax systems. In Section III, we offer an explanation as to why these proposals have gathered such support, despite their limitations, and draw insights from sociology of technology and behavioural science. We conclude, in Section IV, that: (*i*) whilst automation is indeed likely to create challenges to tax systems in the near future, a new tax on automation is not the most appropriate instrument to address them; and (*ii*) the growing support for such a tax reflects political dynamics, unconscious and institutional biases, rather than sound taxation principles, such as equity or efficiency.

2 Why Taxing Robots is Problematic

Whilst the societal changes brought about by the digitalisation and automation of the economy are such that they affect all areas of the tax system, until recently most of the policy attention had been on their impact on corporate income taxes and –to a lesser extent– consumption taxes. This has been progressively changing over the last few years, with the realisation that ongoing labour changes, brought about by automation and the spread of remote working, are also likely to present very significant challenges to personal income taxes [20].

It is hard to overestimate the importance of personal income taxes to tax systems globally. From a revenue perspective, in 2018 a combination of various labour taxes amounted to approximately 50 per cent of total tax revenues in OECD countries, and nearly 60 percent in European countries [21]. From an equity perspective also, personal income taxes are the most important instrument of tax progressivity worldwide, so that any decrease in relative importance of those taxes will have a negative impact on the overall progressivity of the tax system [22, 23].

Given the relevance of these taxes to our global tax systems, it is unsurprising that ongoing labour changes have raised significant concerns from a tax perspective. Whilst the magnitude and distribution of these changes may not yet be known, what is certain is that they have the potential to destabilise what is arguably the most important tax instrument worldwide, with wide economic and societal ramifications.

It is against this background that proposals for taxation of robots have emerged. Although there are significant variations on the basic design of the proposed taxes [14, 24, 25], as previously mentioned they share common aims: to act as a proxy tax for revenue loss; and/or, to function as a regulatory tax to slow down the process of automation. Either way a new tax on robots would be problematic. As a proxy tax, it is unlikely to be an effective instrument to achieve its aim. As a regulatory tax, it is likely to be an effective instrument to achieve its goal, but the goal itself is problematic. Both as a proxy and as a regulatory tax, a new tax on robots or their use would create many conceptual, economic and legal challenges.

First, until robots have legal capacity, taxing robots can only amount to taxing the use of robots [26]. This has various implications, not least the fact that, regardless of its aim, this tax would not be a proxy personal income tax, but a *de facto* tax on capital, i.e. an additional corporate tax. Second, insofar as a tax on robots may function as a regulatory tax to slow down the diffusion of robotics, there are again concerns. Slowing down comes at the expense of the creation of broader economic inefficiencies [27], and large welfare losses [28]. Indeed, once robots become cheaper, or current low- skills workers retire, the gains achieved from taxing robots differently to other capital equipment are close to zero [29, 30]. Third, the regulatory aim of any tax on robots is in itself problematic. Whilst there are clear immediate negative externalities to increased automation, it is also true that it would negate the productivity and welfare gains that would be made in addressing future global challenges such as demographic changes [31] - taxing the use of robots could in effect amount to "shooting one-self in the foot" [32]. Fourth, the 'superstar effect' provides a clear indication of the risks, from an international competition perspective, of introducing such a tax: in the absence of international coordination, the imposition of and additional tax burden on the use of robots would most likely reduce the competitiveness of companies located in that jurisdiction [24]. Finally, conceptually there would be enormous problems in designing such a tax on the income of robots. Many legal design issues arise [14], but from the outset defining what constitutes a robot for the purposes of this new tax, given the difficulty of measuring the work that has been replaced by the robot [33], is not only problematic, but a nearly insurmountable task [17, 24].

Given the significant challenges that a new robot tax would raise, the question that arises is why the idea has nevertheless gained such traction.

3 Why Taxing Robots is Appealing

The choice of a robot tax as an instrument to address the challenges posed by automation may feel intuitive [34], but on closer analysis, it is far from obvious. It can be better explained through the lenses of behavioural science, than that of taxation principles, such as efficiency or equity.

There is now growing evidence that tax policy choices and perceptions are influenced by a variety of institutional biases, as well as cognitive biases and heuristics [35]. Proposals for a new tax on robots may be hard to justify from a legal and economic basis, but is easier to understand when consideration is given to those biases.

First, as a behavioural regulating tool, the question that arises –as with other proposals for regulatory taxes– is whether tax policy is the most appropriate instrument to address the aim that has been set [36]. However, whilst there is theoretical evidence that a robot tax would be an effective way of slowing the pace of automation [28, 29], there is limited evidence to indicate that –even if desirable– that this could be achieved in a competitive setting, in the absence of global tax coordination.

As a substitution tool, to compensate for the decrease in personal income taxes' revenues and decrease in tax progressivity, the key question is whether a proxy tax is the most appropriate instrument to address that aim. Arguably, the most appropriate benchmark for assessing revenue collection levels or progressivity is either the overall tax system –the tax mix– or even the public finance mix, i.e. the tax and expenditure systems put together. Different taxes have different aims, and it is the composition of that tax mix that will determine revenue levels, and the impact of the tax system on inequality [23]. Yet, this is not how we tend to think about revenue collections or progressivity. In practice, much of the analysis –as well as public perceptions– focusses on individual taxes, tax-per-tax, rather than on the tax mix. This is largely a result of a framing heuristic.

Individuals tend to focus on the component of the tax system that they are asked to consider, and fail to consider information on other components of the tax system [37]: thus if the problem is framed as a personal income tax problem, the tendency will be to present a solution to resolve that problem, rather than a solution to resolve the decrease of revenue or progressivity problem. This is coupled with what has been identified as a tax disaggregation bias: an extension of the mental accounting heuristic [38, 39], according to which individuals have an intuition about what tax fairness ought to look like, and will judge the fairness of each tax individually, failing to take into account other elements of their tax system in their appraisal [37].

Second, tax policy design can often be attributable to a fundamental public finance paradox, often expressed by reference to a sentence attributed to US Senator Russel Long: "Don't tax you. Don't tax me. Tax that fellow behind the tree!" [40]. Whilst views on increasing the role of taxation on tackling inequality and redistributing wealth can be widely held [41], those views often translate in a wish that "others" pay taxes – someone else who is not like us [40]. Robots are, of course, not only the ultimate others, but are particularly cool others [42].

Third, as others, robots spark ethnocentric feelings [43, 44]. From the early days of cinema, and science-fiction literature, popular representations of robots have tended to anthropomorphise them in the context of a dystopian future [45]. From the 1980s onwards in particular, in films such as the *Terminator* (1984) or the *Matrix* (1999), robots are consistently represented as an existential threat to humans. Whilst generally popular representations are known to activate ethnocentric views [46], in the absence of anti-racism social norms, the othering of robots, and the manifestation of intra-group bias in relation to them, presents itself as socially acceptable. There is no shortage of historical examples of taxation as a manifestation of ethnocentric feelings [40], and in that context, taxation of robots can be seen also as a protectionist tool: taxes as our last resource against the rise of Skynet and human annihilation.

Finally, institutional biases are likely to play a role. The ongoing labour developments as a result of automation generate uncertainty that is likely to be experienced asymmetrically between losers and winners: while losses are easily identified, gains are more uncertain [47], not least because there are no guarantees they will indeed take place. Therefore, whilst in the long-term robotics and automation may have positive effects on labour demand or productivity, in the short-term, the displacement effects –and the negative effects they carry– are more acutely felt. There is therefore a natural institutional bias towards the status quo, which can be manifested through an attempt to slow down the pace of change, and maintenance of existing (proxy) revenue streams, via the introduction of a new tax on robots. Dependency of existing levels of personal income tax revenues [20] is also likely to enhance this natural bias, as are legal entrenchment [48, 49] and sunk cost fallacy.

4 Conclusion

The debate over the economic and societal consequences of labour changes brought about by automation and digitalisation of the economy is ongoing. Whilst these is still much that we do not know [49], it is now clear however that in the short term those changes present significant challenges to tax systems worldwide. These challenges will be felt particularly by countries –like those within the OECD or the EU– which rely heavily on personal income taxes, both as revenue generators and as key instruments to ensure the progressivity of tax systems.

Tackling them, however, is unlikely to be resolved by the deceivingly simple introduction of a regulatory proxy tax on robots [50]; and rather more likely to necessitate the adoption of a multi-pronged approach. The choice of a particular tax instrument to tackle these challenges – or indeed any others – should be preceded by a two part-test suitability, as follows: *(i)* is tax policy the best way to address the challenges in question?; and *(ii)* if the answer is yes, then what is the best tax instrument to address them. It does not necessarily follow that personal income tax problems, should have personal income tax solutions. Tax policy discussions tend to be framed primarily in terms of key taxation principles, such as efficiency and equity. The proposal for taxation of robots –or their use– are yet further evidence that tax policy discussions are often determined more by unconscious and institutional biases, than those taxation principles. Acknowledging this reality will –perhaps rather paradoxically– result in more efficient and equitable tax systems.

References

1. Burdry Carbo, A., et al.: Law Revolution des Avocats 2.0. Le Temps, 2 December 2016
2. Benmelech, E., Zator, M.: Robots and Firm Investment. NBER Working Paper 29676 (2021). Despite a few dissenting voices
3. Dauth, W., et al.: The adjustment of labour markets to robots. J. Eur. Econ. Assoc. **19**(6), 3104–3153 (2021)
4. Eggleston, K., Lee, Y.S., Iizuka, T.: Robots and labor in the service sector: evidence from nursing homes. NBER Working Paper 28322 (2021)
5. Acemoglu, D., Restrepo, P.: Automation and new tasks: how technology displaces and reinstates labor. J. Econ. Perspect. **33**(2), 3–30 (2019)
6. Acemoglu, D., Legarge, C., Restrepo, P.: Competing with robots: firm-level evidence from France. In: American Economic Association Papers and Proceedings, vol. 110, pp. 383–388 (2020)
7. Graetz, G., Michaels, G.: Robots at work. Rev. Econ. Stat. **100**(5), 753–768 (2018)
8. Aghion, P., et al.: Artificial intelligence, growth and employment: the role of policy. Economie et Statistique **510–512**, 149–164 (2019)
9. Autor, D., et al.: The fall of labour share and the rise of the superstar firms. Q. J. Econ. **135**(2), 645–709 (2019)
10. Dixon, J., et al.: The Robot Revolution: Managerial and Employment Consequences for Firms. NYU Stern School of Business (2020)
11. Boyd, R., Holton, R.: Technology, innovation, employment and power: does robotics and artificial intelligence really mean social transformation? J. Sociol. **54**(3), 331–345 (2018)
12. Carr, N.: The Glass Cage. Norton, New York (2014)
13. Baldwin, R.: The Globotics Upheaval – Globalization, Robotics and the Future of Work. Weidenfeld & Nicolson (2019)
14. Lexer, M.G., Scarcella, L.: Artificial intelligence and labor markets. A critical analysis of solution models from a tax law and social security law perspective. Rivista Italiana di Informatica e Diritto **1**, 54–73 (2019)
15. Delaney, K.: The Robot that Takes Your Job Should Pay Taxes, Says Bill Gates. Quartz, 17 February 2017
16. European Parliament, Report with Recommendations to the Commission on Civil Law Rules on Robotics (2015/2103(INL)), 27 January 2017
17. Falcao, T.: Should my dishwater pay a robot tax? Tax Notes International 1273, 11 June 2018
18. Baldwin, R.: The Globotics Upheaval – Globalization, Robotics and the Future of Work. Weidenfeld & Nicolson (2019). at 11
19. Abbot, R., Bodenschneider, B.: Should robots pay taxes? Tax policy in the age of automation. Harvard Law Policy Rev. **12**, 145 (2017)
20. de la Feria, R., Maffini, G.: The Impact of digitalisation on personal income taxes. Br. Tax Re. **2**, 154–168 (2021)
21. OECD: OECD Global Revenue Statistics Database. https://www.oecd.org/tax/tax-policy/global-revenue-statistics-database.htm
22. Iosifidi, M., Mylonidis, N.: Relative effective taxation and income inequality: evidence from OECD countries. J. Eur. Soc. Policy **27**(1), 57–76 (2017)

23. Akgun, D., et al.: The Effects of the Tax Mix on Inequality and Growth. OECD Working Paper 79 (2017)
24. Englisch, J.: Digitalisation and the future of national tax systems: taxing robots. In: Haslenher, W., et al. (eds.) Tax and the Digital Economy, pp. 261–282. Kluwer Law International (2019)
25. Chand, V., Kostic, S., Reis, A.: Taxing artificial intelligence and robots: critical assessment of potential policy solutions and recommendations for alternative approaches. World Tax J. **12**(4) (2020)
26. Oberson, X.: Taxing robots? From the emergence of an electronic ability to pay to a tax on robots or the use of robots. World Tax J. **9**(3) (2017)
27. Staubhear, T.: On the economics of universal basic income. Intereconomics **2**, 74 (2017)
28. Humlum, A.: Robot adoption and labor market dynamics. Princeton Working Paper (2019)
29. Thuemmel, U.: Optimal taxation of robots. CESifo Working Paper Series 7317 (2018)
30. Guerreiro, J., Rebelo, S., Teles, P.: Should robots be taxed? NBER Working Paper 23806 (2019)
31. Eggleston, K., et al.: Robots and labor in the service sector: firm-level evidence from nursing homes. NBER Working Paper 28322 (2021)
32. Mann, R.: I robot: U tax? Considering the tax policy implications of automation. McGill Law J. **64**(4), 763–807 (2019)
33. García Novoa, C.: Impuestos atípicos en la era post BEPS. In: Cubero Truyo, A. (ed.) Tributos asistemáticos del ordenamiento vigente, pp. 223–224. Tirant lo Blanch, Valencia (2018)
34. Mercier, H.: Not Born Yesterday – The Science of Who We Trust and What We Believe. Princeton University Press (2020). On the power of intuition in decision-making
35. de la Feria, R., Walpole, M.: The impact of public perceptions on general consumption taxes. Br. Tax Rev. **67**(5), 637–669 (2020). Literature cited therein
36. de la Feria, R., Ness, R.: The EU financial transaction tax as an unsuitable and unnecessary proxy tax. Canadian Tax J. **64**(2), 373 (2016)
37. McCaffery, E., Baron, J.: The humpty dumpty blues: disaggregation bias in the evaluation of tax systems. Organ. Behav. Hum. Decis. Process. **91**, 230–242 (2003)
38. Thaler, R.: Mental accounting and consumer choice. Mark. Sci. **4**, 199–214 (1985)
39. Thaler, R.: Mental accounting matters. J. Behav. Decis. Mak. **12**, 183–206 (1999)
40. Keen, M., Slemrod, J.: Rebellion, Rascals and Revenue. Princeton University Press (2021). at 132
41. Stantcheva, S.: Understanding tax policy: how do people reason? NBER Working Papers 27699 (2020)
42. Singer, P.: Wired for War – The Robotics Revolution and Conflict in the 21st Century. The Penguin Press, New York (2009). at 1
43. Baldwin, R.: The Globotics Upheaval – Globalization, Robotics and the Future of Work. Weidenfeld & Nicolson (2019). at 8 et seq. On the instrumentalisation of ethnocentric feelings generally
44. Sobolewska, M., Ford, R.: Brexit Land, pp. 85–118. Cambridge University Press (2020)
45. The etymology of the word itself – from the Czech for 'forced labour'– points towards an anthropomorphised view of AI
46. Ang, D.: The birth of a nation: media and racial hate. Harvard Kennedy School Working Paper No. RWP20-038 (2020)
47. Fernandes, R., Rodrik, D.: Resistance to reform: status quo bias in the presence of individual uncertainty. Am. Econ. Rev. **81**, 1146 (1991)
48. Bell, J.: Path dependence and legal development. Tulane Law Rev. **87**, 787 (2012)
49. Ranchordas, S.: One foot in the door: evidence- based limits on the legislative mandate. Hukim - J. Legislation **207** (2018)
50. Sunstein, C.: Simpler: The Future of Government. Simon & Schuster (2015). On the appeal of simple solutions in public policy

Taxing Robots or Artificial Intelligence

Xavier Oberson[1,2](✉)

[1] University of Geneva, Geneva, Switzerland
xoberson@obersonabels.com
[2] Oberson Abels SA, Geneva, Switzerland

Abstract. This paper intends to replace the debate about the current situation with regards to the taxation of robots and to summarize the main legal arguments in favor of a robot tax. In particular, since so far robots as such do not have an economic ability to pay, at least under the current state of law, the justification of a robot tax should clearly distinguish between taxing the use of robots on the one hand, and robots as such on the other hand. A more recent argument in favor of taxing robots as such has also emerged with a view to keep the integrity of the tax system and to fight against tax avoidance and fraud. In any event, the implementation of such proposal would require an international coordination.

1 Introduction

The idea of taxing robots has emerged around 2016. Such a concept has been studied at the EU level, notably in an important report from the EU Parliament[1], and has gained worldwide attention, following a short interview of Bill Gates who, openly, favoured such a proposal[2]. On our side, we also have suggested to tax robots, under specific conditions[3], and have tried to present various options in favour of such a system of taxation, going even further by considering not only to tax the use of robots, but taxing

[1] See, notably, European Parliament, Committee on Legal Affairs, Draft Report with recommendations to the Commission on Civil Law Rules on Robotics (2015/2103(INL)) 2016, p. 10.
[2] Gates B., Interview with Quartz, 17th February 2017.
[3] Oberson X., Taxing Robots? From the Emergence of an Electronic Ability to Pay to a Tax on Robots or the Use of Robots, 9 World Tax Journal 2017, p. 247 ff, 251; Oberson X., "Taxer les robots?", Bilan, 5th July 2016; Oberson X., "Un jour les robots refuseront de payer leurs impôts!", Le Temps, 17th October 2016.

M. A. Grau Ruiz (Ed.): INBOTS 2021, BIOSYSROB 30, pp. 100–105, 2022.
https://doi.org/10.1007/978-3-031-04305-5_17

robots as such[4]. Other commentators have also developed or analysed further this idea, in an interesting and lively debate[5].

Eventually, this concept has gained much attention among policy makers. Many authors, probably still the majority view today, reject that proposal for various reasons, which, without going into details, either focus on the economic impact of automation, or on the legal difficulties of implementing a robot tax. Indeed, from an economic standpoint, under the so-called optimist view, artificial intelligence (AI), combined with robots, should eventually create many new jobs, which would clearly not justify the introduction of a new tax on robots. In addition, taxing robots would discriminate innovation. The legal criticism focuses on the difficulty, if not impossibility, to define a robot for tax purposes, and the problems to implement such a tax in an efficient and adequate form.

We have tried to answer some of these criticisms elsewhere[6]. What we would like to address in this short presentation is to: (i) replace the debate in its current situation and correct some potential misunderstanding of the proposal; (ii) summarize the main legal arguments in favour of a robot tax, and (iii) plead in favour of an international coordination.

2 Current Perspective of the Proposal

From an economic standpoint, the optimists, based on a Schumpeterian vision of the famous concept of "creative destruction", argue that AI would certainly destroy many jobs, but new jobs would be created and eventually even more jobs than previously, contributing to a collaboration between humans and robots. Many prominent economists and analysts currently share this view. We should however also listen to the opposite standpoint. From the pessimist's sides, AI is designed to replace, in the long term, not only human work within factories and industries, but also to challenge the brains of humans. Eventually, any jobs could be replaced by a machine with an AI.

In our view, we don't know who is right or wrong in this debate. However, we still believe that many jobs will disappear because of the development of AI. As such, this is not a problematic aspect, as long as boring or even dangerous jobs are replaced. But in

4 Oberson X., Taxing Robots, Helping the Economy to Adapt to the Use of Artificial Intelligence, Elgar Publishing, Cheltenham, UK, Northampton MA, USA, 2019, p. 9 ff.

5 See, among others, Abbott R./Bogenschneider B., Should Robots Pay Taxes? Tax Policy in the Age of Automation, 12 Harvard Law & Policy Review 2018, p. 1 ff; Mazur O., Taxing the robots, 46 Pepperdine Law Review 2018, p. 1 ff.; Englisch J., Digitalization and the Future of National Tax Systems: Taxing Robots?, in: Haslehner W./Kofler G./Pantazatou K./Rust A. (ed.), Tax and the Digital Economy, Wolters Kluwer, The Netherlands 2019, p. 261 ff.; Kovalcev R., A Taxing Dilemma: Robot Taxes and the Challenges of Effective Taxation of AI, Automation and Robotics in the Fourth Industrial Revolution, The Ohio State Technology Law Journal 2020, p. 182 ff; Bogenschneider B., Will Robots Agree to Pay Taxes? Further Tax Implications of Advanced AI, North Carolina Journal of Law & Technology, vol. 22, 2020, p. 1 ff.

6 See, recently, Oberson X., Robot taxes: The rise of a new taxpayer, Bulletin for international taxation 2021, p. 370 ff.

the long run, the scope of jobs affected by AI would constantly expand and, potentially, all sectors, notably services, could also be affected[7]. At least, many workers would not be able to adapt rapidly enough to the disruption caused by automation. Should the pessimists be right, the States would be in a very delicate situation, notably for the financing of the social security system and the economy as a whole. Indeed, should labour income shrink, revenues from all taxes and social security levies applicable to salaries would drastically be reduced. Our point here is to adopt a precautionary approach, which is to already analyse and design practicable models of taxation of AI or robots that could be introduced, should eventually the impact of automation of labour meet the predictions of the pessimists. The growing inequality in the distribution of wealth among people, linked with the automation of the economy is also a factor of concern that a robot tax could address. The complexities and difficulties to implement global rules of taxation of multinationals entities, notably in the digital economy, leading in particular to the adoption of the so-called pillar one and two proposals, under the Inclusive Framework at the OECD level[8], have demonstrated how time consuming and uncertain it is to reach an international consensus in such major tax policy issues.

3 Legal Justifications of a Tax on Artificial Intelligence or Robots

In order to face the impact of AI on labour, should the pessimists be right, there are various solutions. One of them, which the author has been advocating since many years, is taxing robots. How do we justify then the taxation of robots from a legal standpoint? Here the discussion should focus on the admissibility of such a tax, from an economic and constitutional standpoint. Indeed, there are various potential forms of taxing robots and each of them requires a different legal and economic justification. For instance, should we consider robot taxes, like ecotaxes, as some form of Pigouvian taxes designed to internalize the social costs linked with unemployment, the legal justification should analyse the compatibility of such goal and the proportionality of its implementation through a tax. By contrast, a general tax on robots, targeting the so-called "imputed salary" of the use of robots by enterprises, should focus more on its compatibility with the global tax system, try to avoid double economic taxation and be fair and equitable. The same kind of legal analysis should also apply to so-called proposal for an automation tax. This tax would focus more on the production factors, based on a ratio, such as the number of human workers in comparison to the profit, or to the turnover of the company. The impact of these taxes would require a delicate analysis both of its economic impact and from an equality of treatment standpoint. Finally, a tax on robots as such, requires a

[7] In a recent study, it was shown that pension funds managed by AI could be more successful than the one managed by humans, see Garessus E., "Une intelligence artificielle bat largement les indices des caisses de pensions", Le Temps, 26th January 2022.

[8] OECD, Brochure, Two-Pillar Solution to Address the Tax Challenges Arising from the Digitalisation of the Economy, October 2021; OECD, Addressing the tax challenges arising from the digitalization of the economy, July 2021.

more delicate legal analysis because it requires notably to allocate some kind of ability to pay to these entities called "robots", to the extent we may define them.

We have demonstrated earlier that, from a legal standpoint, a different analysis should be made between a tax on the use of robots, on the one hand, and on robots as such, on the other hand. Indeed, despite some proposals, notably under the EU Parliament report, robots so far do not have legal personality. As such, they do not have any ability to pay the tax, namely the power to legally control the attributions of the funds received and transferred to third party. Thus, we have advocated to work on two distinct phases. First, we should focus on a tax *on the use* of robots. This could justify a tax on the so-called "imputed salary" on the use of robots by enterprises, namely an estimate of the wages that enterprises would have paid for the use of humans instead of robots. At the VAT level, a tax on the use of robots is already applied and could be even more developed. Indeed, a lawyer, having used the services of the robot Ross, a legal computer specialized in legal research, would include in his or her legal bills, potentially subjects to VAT, the cost of the use of the robot Ross to perform the legal service to the final customer. When an enterprise, subject to VAT, sells a car which has been built using automation, the price of the car sold includes the VAT on the costs, which have been provided by automation.

In parallel, in a more traditional perspective, there also is the possibility to tax robots as *objects*, like a tax on cars, boats or dogs. In fact, this kind of taxes already exists today. It comes to no surprises that the first implementation of robot's taxes around the world have taken this form. For example, some States in America have implemented taxes on self-driving cars or drones[9]. In Switzerland, the Canton of Geneva has also proposed a tax on self-vending machines, which has not been accepted so far.

However, we can even imagine that, at the second stage, in the future, if a legislator would start granting a legal personality to a robot, a r*obot as such* could be subject to tax. This would of course require that robots, as defined by the legislator, would benefit from an ability to pay. However, in the case of robots, we are not considering, for our purpose, a subjective ability to pay, like humans, but we focus on a *functional ability to pay* in the form of a capacity of payment. This approach would be similar to corporation taxes, which are justified by the separate entity approach. In addition, in order for an entity to be subject to tax, it is not even necessary to be recognized as a legal person, as it is already the case in current tax law[10], but sufficient to have a capacity of payment, in the form of a separate patrimony, or at least specific assets legally attributed to the robots as such[11].

---

[9] See the examples from Kovacev, *supra* note 5 (2020), p. 203 ff.

[10] Oberson, *supra* note 3 (2017), p. 251; same opinion, Dimitropoulou Ch., Robot Taxes: Where Do We stand? In: CFE 60th Anniversary, Liber Amicorum, IBFD 2019, p. 55 ff, 65.

[11] Oberson, *supra* note 6 (2021), p. 375; Englisch, *supra* note 5 (2019), p. 264.

The author is aware that such an approach is rather controversial. To be clear, this vision has nothing to do with an automorphism view or robot. We don't suggest a tax on robots because "we like robots". We consider the legal concept of robot from a functional perspective. The definition of robots, as new entities subject to tax should be based on criteria, which, from a tax perspective, focus on their impact and functions on the labour market. In fact, when you think about it, let us say more than one century ago, we were confronted with a similar situation. Indeed, some States started to suggest that corporations, under some specific characteristics, could become new legal entities, with legal rights and obligations. As such, corporations are just legal creations of a legislator. To exist as a legal entity, a corporation must meet the legal requirements fixed by the law, typically a minimum amount of equity, a goal, a board of directors, and sufficient formal rules of accounting and registration. Perhaps, are we now confronted with the same issue and are faced with the need to define a new form of legal or digital entity such as AI or robots. We do not see why a legislator could not find an appropriate definition of robots or AI subject to tax.

However, the concept of robots for tax purposes should not be based on its physical form, but a "form neutral" definition of robots is required. What is relevant is the use of AI and its impact on the labour market. The activities of robots are indeed done through algorithms that are embodied either in a physical or digital form. We have advocated to focus on "smart robots" (or smart algorithms) which are autonomous and have the capacity to interact with third parties and learn. Should the robots become legal persons, or at least tax subjects, we could then try to implement specific taxes on them. The evolution of technology, via Blockchain or crypto currencies, could eventually event help to allocate a financial capacity to robots and grant them the rights to levy a tax on transfer of payments. A more revolutionary approach could also consider that robots, as such, would be regarded as subjects to VAT. They could then charge output VAT on their supplies and credit input VAT on their purchases.

In addition, a more recent additional argument in favour of taxing robots as such could eventually focus on the need to keep the integrity of the tax system, with a view to fight against tax avoidance and fraud[12]. Some publications in the United States have demonstrated that, already under current law, an algorithm could take control of an LLC and thus get the possibility to enter into the market with potentially harmful results[13]. This triggers a risk of tax avoidance or fraud. Such a concern would be compounded, should, at a later stage, the algorithms, through the use of automated AI, enter into transactions, without sufficient control from any human beings, or entities, capable as serving as taxpayers, which entails a further risk of income not being subject to tax, not only in the absence of a taxpayer, but also without an appropriate nexus to allocate the relevant profits to it. This development would also favor the introduction of robots, as defined by legislator, being subject to tax[14]. In a recent interesting article, Bogenschneider analyses

[12] Oberson, *supra* note 6 (2021), p. 377; De Lima Carvalho L., Spiritus Ex Machina: Addressing the Unique BEPS Issues of Autonomous Artificial Intelligence by Using "Personality" and "Residence", 47 Intertax 2019, p. 425 ff.

[13] De Lima Carvalho, *supra* note 12 (2019), p. 425 ff.

[14] Oberson, *supra* note 6 (2021), p. 380.

the further implications of an advance AI, which could soon have the ability to engage in "factual structuring" in order to avoid tax[15].

4 A Need for an International Coordination

Finally, it makes however little sense to develop uncoordinated and controversial robot taxes at the domestic level. AI and robots are easy to be moved, via servers or in the cloud, in favourable tax jurisdictions. The EU could already be a right place to start analysing the essential features of a robot tax[16]. In any event, an international coordination at the OECD or UN level would appear necessary. As the discussion on the pillars 1 and 2 solution have demonstrated, it takes a long time to find a consensus on such new and global ideas of taxation. Therefore, the author believes that we should study the feasibility of a tax on robots right now.

---

[15] Bogenschneider, *supra* note 5 (2020), p. 10 ff, who argues that, as AI will become well informed about the economy, advanced AI will demand a type of income taxation levied at high rates, both on themselves and on humans, *id*. p. 43 ff. This conclusion seems debatable, since it would require also that, in parallel to the evolution of AI, the tax system would remain based on the same principles. An analysis of this potential development is however outside the scope of our contribution.

[16] See Dimitropoulou, *supra* note 10 (2019), p. 65 ss, who suggests at least a tax at the EU level.

Taxes on the Use of Robots

Álvaro Falcón Pulido(✉)

Law, Complutense University of Madrid, Madrid, Spain
alfalcon@ucm.es

Abstract. One of the most novel issues to be faced by tax law is the impact of robotics in today's society. This raises several possibilities: creating a tax on the robots themselves for the income generated; creating a tax on the income generated through robots that falls on their owners, or a tax on the use of robots, which is what we briefly analyse in this contribution.

1 Introduction

There are several possible solutions to the social and economic problems [1–3] that the intensive and extensive development of robots may bring about. They include the adaptation of existing taxes, the creation of specific taxes, or even a combination of both.

Efforts should be focused on improving existing taxes and those to be created, to build a more harmonious and fluid tax system. We do not consider that earmarking [4] new taxes is just the right path to follow.

In some cases, it may be possible to reduce the taxation of robots to the payment of a simple fee [5, 6] for the development of an administrative activity, such as obtaining a license plate or a mandatory inspection, as in the Spanish Vehicle Technical Inspection. However, it would also be acceptable for this fee to be complemented, where appropriate, with other taxes to try to better accommodate the new robotic reality.

The creation of a tax on robots has been proposed from various quarters in recent years. Exactly what this tax means and what would be its taxable event, its taxable person [7], its taxable base, etc. varies greatly depending on the proposals. Most of them do not go into these details in a specific way, being more a recommendation for the future and consideration of the possibilities.

In the following sections we will analyse some of the legal characteristics of a possible tax on robots, especially the typology and structure. Our analysis will be focused on a tax on the use of robots and not on a tax on robots (per se) [8], i.e., the analysis will be focused on the basis that robots are the economic object of the tax and not the subject of the tax.

2 Different Types of Taxation

In order to establish a new tax on robots, we must first consider what type of tax we want to implement. The classification of taxes into different typologies is very varied. They can be divided with respect to multiple criteria, several of which are not particularly relevant

M. A. Grau Ruiz (Ed.): INBOTS 2021, BIOSYSROB 30, pp. 106–110, 2022.
https://doi.org/10.1007/978-3-031-04305-5_26

in the case of robotics, such as periodic and instantaneous taxes, which we will not discuss here (although automation is increasingly providing real time tax management). Three of the commonly used classifications may be relevant in determining the taxation of robots and will therefore be discussed below.

2.1 Direct and Indirect Taxation

"Direct taxes are those that are levied on the basis of a direct index of economic capacity" such as income or wealth, while "indirect taxes are those that are based on an indirect index" [9–12] such as consumption. Most proposals on the creation of a tax on robots are direct taxes, since they are based on the income generated by robots, whether real or imputed. It seems obvious that if a tax on robots (being their taxable subject) or on the use of robots were to be created, it would most likely be a direct tax.

However, it cannot be ignored that robots are already subject to indirect taxes, such as VAT, like any other good that is the subject of sales and purchases. And even the sale of software can be considered a service for VAT purposes, if certain requirements are met. Nor can excise duties and other indirect taxes be ignored.

2.2 Subjective and Objective Taxation

In subjective taxes the tax regulation takes "into account the personal circumstances of each taxpayer, adapting the burden to such circumstances" [13–15] while in objective taxes the tax regulation ignores, or does not take into account, the personal circumstances of the taxpayer. In the case of establishing a tax on robots it seems that it would be in most cases a subjective tax, either because it would be a tax that takes into account the characteristics of the owner of the robot, or of the robot itself, in the case that this would be its taxable subject with a robotic personality.

However, robots are perfect candidates for an objective tax to be levied on them [16], similar to the Real Estate Tax. Such a tax has its social justice and equity shortcomings that would force it to be a relatively low-impact tax with respect to revenue. Even so, it is likely that the first taxes to be implemented on robots will be of this type, due to their simplicity of application, implementation and review.

We have to recall again the case of VAT, which is an objective tax and affects robots at present, and would affect them in case they were robotic persons acting in legal-economic traffic.

2.3 *Personal and Real Taxation*

"This distinction [...] is established by a doctrine (SAINZ DE BUJANDA) generally accepted in the following terms: personal taxes are those whose taxable event cannot be thought of except in relation to certain persons; otherwise, taxes are considered real" [17–20].

In the field of robotics, it seems reasonable to expect that most taxes would be of a personal nature, since most proposals focus on income. However, it is worth noting that it is possible to establish a real tax on robots, similar to the Real Estate Tax or the Motor

Vehicle Tax. A tax of this type seems reasonably straightforward, saving the difficulty of defining a robot, and it is possibly an easy tax to implement, at least on a specific portion of robots, such as autonomous vehicles, drones or domestic service robots.

3 Structure of the Robot Tax

It is not possible to analyse the structure of the tax on robots in detail as it does not exist yet, and there are no proposals for specific laws. The possibilities of creating taxes on robots vary as much as the possibilities of creating new taxable events to be taxed. However, it is possible to analyse some general aspects of the elements of the taxable event of possible taxes on robots or around robotics that have a special consideration because of the subject matter involved, robotics.

3.1 Subjective Element

The subjective element is "the relationship, in fact or in law, in which the person liable to pay the tax must stand in relation to the material element of the tax" [21–23]. In the case of robotics, we can clearly differentiate two cases, the robot as a taxable person and the robot as part of the material element. Of course, "the subjective element should not be confused with the taxable person", the subjective element is a relationship while the taxable person is the taxpayer, or his/her substitute [24].

Robots could only be taxable persons, as we have seen in chapter two, if they had a specific personality. In this case, and for example in the case of an income tax, the robot would be the taxpayer who obtains the income. Whereas, in the case of a tax on income from the use of robots, the robot would be a part of the objective element necessary for the income obtained by the owner or user of the robot to fulfil the subjective element.

3.2 Objective Element

The objective element or presupposition consists of the material, spatial and temporal aspects of the taxable event [25] and the material element "comes to coincide with the wealth or manifestation of economic capacity embodied" [26, 27] in the taxable event. In the case of robotics there do not seem to be any specificities related to the nature of robotics and in general the objective elements of the taxable event will coincide, or at least be similar, to those of other taxes similar to those created on robots.

4 The Concept of Notional Imputed Income

One of the biggest problems facing a tax on the use of robots is how to determine the value of that use. This is not a tax on robots as taxable persons with personality, in which case it would be relatively straightforward to determine the income they generate, but a tax that seeks to tax the income generated by the use of the robot.

If robots replace workers, the income of these workers is no longer taxed and is not replaced by a "real" income from the robots that can be taxed and therefore a way must be found to "bring back" this income into the tax system [28]. To remedy this loss of revenue through a tax on the use of robots, some authors have recommended the introduction of a "imputed salary" [28], i.e., a theoretical imputed income. This would make the tax treatment of robots and workers neutral, as owners would have to take into account both the workers' wages and the theoretical wages of the robots, which would be taxed at the owner's level [28].

In Spain there is already a situation that could be considered similar, the imputation of real estate income in the personal income tax. Just as this potential income of property owners is taxed, the potential savings of workers' income, or, in other words, the potential salary of robots, could be taxed.

Of course, this solution faces serious practical problems. How we determine this theoretical imputed rent correctly is a difficult problem to solve, if it can be solved at all. Robots can partially replace workers, or replace tasks of different workers [29]. It may also be the case that they require new workers to work alongside them. They may replace workers of different skills and wage levels. And they may not replace any workers by performing tasks that could not be done before.

The one way to carry out a correct and fair determination of the income to be imputed would be a detailed study of the functional analysis, similar to the transfer pricing methodology [29].

Other solutions to this problem are based on ratios, such as a ratio between workers' wages and company profits as proposed by Abbott/Bogenschneider. However, these proposals also have their complications. From which ratio should be applied and how it should be calculated that it would affect different types of industries differently due to their varying degree of robotisation [30].

Acknowledgment. This work was supported by INBOTS CSA Project, Inclusive Robotics for a better Society, Program H2020-ICT-2017–1; Project No. 780073, http://inbots.eu/.

References

1. Grau Ruiz, M.A.: As highlighted by the constant doctrinal preoccupation with this issue in recent years. La adaptación de la fiscalidad ante los retos jurídicos, económicos, éticos y sociales planteados por la robótica. *In:* Nueva fiscalidad, no. 4, pp. 35–61 (2017)
2. Ayala Sanhueza, A.G.: Incentivos fiscales a la innovación en robótica: Una necesaria reformulación. In: Revista general de legislación y jurisprudencia, no. 2, pp. 207–260 (2019). ISSN 0210–8518
3. Quílez Moreno, J.M., Aparicio Chofrégil, L.: Robots e inteligencia artificial: Debería exigirse algún tipo de cotización?. Diario La Ley, Editorial Wolters Kluwer, no. 5, 14 de marzo de 2017 (2017)
4. García Novoa, C.: There are several proposals for making robotics contribute to public spending. Most focus not so much on the structure of taxes on robots, but on the nature of the tax. What is important is not how robots are taxed, but rather to ensure that the revenue is used for specific purposes. In: Cubero Truyo, A. (Dir.) Impuestos atípicos en la era post BEPS (Capítulo 8), Tributos asistemáticos del ordenamiento vigente, Tirant lo Blanch, Valencia, p. 221 (2018)
5. Oberson, X.: Taxing Robots, pp. 130–131. Edward Elgar, Cheltenham (2019)

6. Grau Ruiz, M.A.: Taxing autonomous vehicles: the Californian case. In: Pons, José L. (ed.) INBOTS 2018. BB, vol. 25, pp. 45–50. Springer, Cham (2020). https://doi.org/10.1007/978-3-030-24074-5_10
7. Falcón Pulido, Á.: Taxing robots: clarifications on legal and economic capacity, capacity to act and representation. In: Pons, J.L. (ed.) INBOTS 2018. BB, vol. 25, pp. 40–44. Springer, Cham (2020). https://doi.org/10.1007/978-3-030-24074-5_9
8. Oberson, X.: Taxing Robots, p. 113. Edward Elgar, Cheltenham (2019)
9. Pérez Royo, F.: Derecho Financiero y Tributario. Parte General, Civitas Thomson Reuters, 25° edición, Cizur Menor, Navarra, p. 55 (2015)
10. Falcón y Tella, R.: Derecho Financiero y Tributario (Parte General), Servicio de Publicaciones Facultad de Derecho, Universidad Complutense de Madrid, 8° edición, p. 62 (2019)
11. Martín Queralt, J., Lozano Serrano, C., Tejerizo López, J.M., Casado Ollero, G.: Curso de Derecho Financiero y Tributario, Tecnos, 22° edición, Madrid, pp. 74–75 (2011)
12. Menéndez Moreno, A. (Dir.) Derecho Financiero y Tributario. Parte General, Lex Nova Thomson Reuters, 15 edición, Madrid, pp. 150–151 (2014)
13. Pérez Royo, F.: Derecho Financiero y Tributario. Parte General, Civitas Thomson Reuters, 25 edición, Cizur Menor, Navarra, p. 56 (2015)
14. Falcón y Tella, R.: Derecho Financiero y Tributario (Parte General), Servicio de Publicaciones Facultad de Derecho, Universidad Complutense de Madrid, 8 edición, p. 62 (2019)
15. Martín Queralt, J., Lozano Serrano, C., Tejerizo López, J.M., Casado Ollero, G.: Curso de Derecho Financiero y Tributario, Tecnos, 22 edición, Madrid, p. 73 (2011)
16. Oberson, X.: Taxing Robots, pp. 129–130. Edward Elgar, Cheltenham (2019)
17. Pérez Royo, F.: Derecho Financiero y Tributario. Parte General, Civitas Thomson Reuters, 25 edición, Cizur Menor, Navarra, p. 56 (2015)
18. Falcón y Tella, R.: Derecho Financiero y Tributario (Parte General), Servicio de Publicaciones Facultad de Derecho, Universidad Complutense de Madrid, 8° edición, p. 63 (2019)
19. Martín Queralt, J., Lozano, Serrano, C., Tejerizo López, J.M., Casado Ollero, G.: Curso de Derecho Financiero y Tributario. Tecnos, 22° edición, Madrid, p. 73 (2011)
20. Cazorla Prieto, L.M.: Derecho Financiero y Tributario, Thomson Reuters Aranzadi, 18° edición, Cizur Menor, Navarra, pp. 201–202 (2018)
21. Pérez Royo, F.: Derecho Financiero y Tributario. Parte General, Civitas Thomson Reuters, 25° edición, Cizur Menor, Navarra, p. 151 (2015)
22. Falcón y Tella, R.: Derecho Financiero y Tributario (Parte General), Servicio de Publicaciones Facultad de Derecho, Universidad Complutense de Madrid, 8 edición, p. 165 (2019)
23. Martín Queralt, J., Lozano Serrano, C., Tejerizo López, J.M., Casado Ollero, G.: Curso de Derecho Financiero y Tributario, Tecnos, 22 edición, Madrid, p. 228 (2011)
24. Falcón y Tella, R.: Derecho Financiero y Tributario (Parte General), Servicio de Publicaciones Facultad de Derecho, Universidad Complutense de Madrid, 8° edición, pp. 165–166 (2019)
25. Falcón y Tella, R.: Derecho Financiero y Tributario (Parte General), Servicio de Publicaciones Facultad de Derecho, Universidad Complutense de Madrid, 8° edición, pp. 162–163 (2019)
26. Pérez Royo, F.: Derecho Financiero y Tributario. Parte General, Civitas Thomson Reuters, 25° edición, Cizur Menor, Navarra, p. 150 (2015)
27. Martín Queralt, J., Lozano Serrano, C., Tejerizo López, J.M., Casado Ollero, G.: Curso de Derecho Financiero y Tributario, Tecnos, 22° edición, Madrid, pp. 227–229 (2011)
28. Oberson, X.: Taxing Robots, p. 114. Edward Elgar, Cheltenham (2019)
29. Oberson, X.: Taxing Robots, p. 115. Edward Elgar, Cheltenham (2019)
30. Oberson, X.: Taxing Robots, pp. 120–121. Edward Elgar, Cheltenham (2019)

The Need to Assign a Tax Value to Economic Activities Based on Data Transactions

J. J. Hinojosa(✉) and G. Sánchez-Archidona

Tax Law at University of Malaga, Malaga, Spain
{jjhinojosa,gsanchezarchidona}@uma.es

Abstract. Companies that base their business model on the "data economy", carrying out operations with data, mostly associated with exploitation, are those that have the largest market share globally, although this does not mean that they contribute accordingly to the public purse. The main problem is that no emphasis is being placed on how to value the intangible asset without which this benefit would not be produced: the data, both raw and refined. Therefore, in this paper, based on the study of the regulatory instruments and published proposals, we will address one of the greatest challenges of tax law: to design precise instruments for measuring the tax value of data, which implies, on the one hand, differentiating between the different types of services provided, and on the other hand, proposing solutions beyond the corsets imposed to date, in order to provide solutions that contribute to existing studies.

1 Introduction

In 2018, data surpassed oil as the most precious commodity on the planet, and there is a small group of companies that have it in their possession and exploit it, in one way or another, for business profit, such as, among others, Google, Amazon, Facebook and Apple (known as GAFA), not always by capturing transparent to the users producing it.

We are referring to multinational enterprises such as the GAFA group: their business models are based on the massive exploitation of data whose economic benefit lies, to a large extent, in the analysis of user data. In other words, these companies are the ones that have best exploited the "data economy" to obtain incessant economic benefits, not always rewarded in terms of contributions to the public coffers of the States.

This is precisely one of the great challenges that tax law faces: to assign a tax value to truly important operations that base their main asset on the analysis and exploitation of data and to modify the system of company-user relations around them. This is the object of study, as well as to propose real and effective solutions, framed within the general objectives of the research project that support this study, of which the signatories are members.

Ministerio de Ciencia, Innovación y Universidades español. RTI2018- 093553B-100: "Legal Tax Challenges of Robotics and Artificial Intelligence in the Digital Era".

M. A. Grau Ruiz (Ed.): INBOTS 2021, BIOSYSROB 30, pp. 111–114, 2022.
https://doi.org/10.1007/978-3-031-04305-5_25

2 Material and Methods

This work has been based On regulatory instruments and proposals at the national or international level that attempt, in some way, to try to quantify data operations. In particular, at a national level, Law 4/2020, of October 15, and at an international level, proposals such as the North American DASHBOARD experience, which is still pending approval.

Thus, the method followed to carry out this research is fundamentally deductive, starting from the instruments or proposals to propose solutions in this regard.

3 Results

The research carried out yields dubious results, still incipient, for regulating data operations, due to the lack of precise regulatory instruments to address this issue. It is clear that, without them, no progress will be made in the taxation of these kind of business.

4 Discussion

The main discussion is to specify the division between operations with problematic data in order to be quantified and those that do not. These operations must be divided into two types: A and B.

Type A: users as active co-producers

The first type refers to situations involving a service provided by a company specializing in data exploitation to a user, entity or contracting Administration. In this case, the data may be either the applicant's own data or data collected by the company in question.

What is really valuable is not the data in isolation or who is the owner or assignee of the data, but the conclusions that these data yield once they have been exploited. These results normally reveal a situation or key data for the contracting party, which in the short, medium or long term will provide substantial economic benefits, much more than the cost of providing the service.

Type B: users as passive co-producers

This type refers to those situations in which some companies obtain information from users, which can subsequently produce economic benefit, either directly, by taking advantage to continue offering products on their platform (Amazon, Aliexpress, Ebay, etc.) or indirectly, if such data is sold to other companies for their own purposes (Facebook).

Actually, it is the one that represents a big problem for tax purposes, because it is not really possible to elucidate: one, how much the (raw) data they obtain from users is worth; and two, what economic benefit they obtain from them after exploiting them for a certain purpose.

5 Review of Existing Proposals

A) *The third taxable event of the Spanish Tax on Certain Digital Services: a bad pioneer experience in the taxation of data transactions.*

The Spanish Digital Services Tax, in force since January 16, 2021, intends to tax the simple transmission of user data, i.e. the transmission of raw data, and these have very little value compared to the same after the refining process.

And if only the "income from data" is to be taxed, the most lucrative aspect is excluded: the process of analysis and exploitation, on the one hand, and the purpose of obtaining the conclusions drawn from the data, on the other.

In short, this third taxable event is not intended to tax the real economic benefit arising from the exploitation of refined data, but only their transmission in raw form, the value of which is substantially lower. It would not solve the problem either of taxing the operation in question or of determining how much the data are worth.

B) *The U.S. DASHBOARD proposal: the right way forward*

The pioneering experience contained in the Designing Accounting Safeguards to Help Broaden Oversight and Regulations on Data Bill (DASHBOARD, SIL 19759), introduced by Senators Warner and Hawley in the US Senate in April 2019, is one of the most interesting paths to explore.

There is no doubt (beyond the path it may take) that it is the proposal that has best understood what is the main focus of conflict raised by the absence of accurate methods of data valuation and the purpose for which they are used.

If the problem is that it is not possible, or the means do not yet exist, to find out how much the data collected by some companies are worth, one solution would be to articulate a binding legal instrument that obliges them to identify them; and thus dispense with the arduous task of evaluating and measuring (also for tax purposes) such data. Something similar is currently happening, following the adoption of the CAD7 Directive and the obligation for intermediary platforms to provide information.

In our opinion, on a theoretical level, it is a very good proposal; however, on a practical level, we have more doubts. Fundamentally it is due to the evaluation methodology of the data obtained, because, although an infinite amount of data can be obtained, it will be of little use to have received them if their assessments are not adequate and correct. Furthermore, it goes without saying that this valuation methodology must be articulated respecting the rights and guarantees of the taxpayers.

6 Conclusions

One of the greatest challenges consists of designing precise methods of quantification of the data that bind the administrations and taxpayers in order to guarantee legal certainty, on the one hand, and, on the other, the creation of the corresponding tax obligations, which will be revealed by taking into consideration the data as a measure of economic capacity.

This means, inevitably, correctly delimiting which types of acts, situations or facts demonstrate this economic capacity and which cannot be included in traditional taxes. In those services whose data are not captured in a conscious manner by the user, or the user does not know about this activity and passively collaborates in the delivery of these data (the passive modality that we have described in this paper), the problem is not framed exclusively in the tax sphere, but in the lack of a legal instrument to regulate this system of relations, which can subsequently generate various tax obligations.

In short, it is intended not only to address a change in the status quo between companies and users around data, but also to establish a new system of relationships between tax administrations and tax payers. And the end of the day, it is about modifying and entrenched system, and it will not be easy at all, nor will it propose new solutions that are accourate in intention bur inefficient in practice (the U.S.A. DASHBOARD, which established a rule for companies to disclose the value of the data collected) nor recycle others not designed to address such a problem (frame the services associated with the exploitation of data in the Spanish DST.

References

1. Hinojosa, J.J.: El difícil camino de la fiscalidad sobre los beneficios de la economía digital en la OCDE. In: Hinojosa, J.J., Cruz, I. (dirs.) Tributos, servicios digitales y robótica, pp. 21–55. Aranzadi, Navarra (2020)
2. Hinojosa, J.J.: Towards a tax on a digital services: an approach from Spain. In: Olesti Rayo, A., Andrés Aucejo, E. (dirs.) Thomson Reuters, Navarra, pp. 213–234 (2021)
3. Hinojosa, J.J.: Una nueva fiscalidad para la economía digital. Gravamen justo o necesario? In: Hinojosa, J.J., Álvarez, J., Tirado, M.C., Cruz, I. (Dirs.) Retos de la sociedad digital. Regulación y fiscalidad en un contexto internacional, Reus, Madrid, pp. 291–334 (2022)
4. Sánchez-Archidona, G.: La tributación de la robótica y la inteligencia artificial como límites del Derecho financiero y tributario. Quincena fiscal **12**, 69–100 (2019)
5. Sánchez-Archidona, G.: El derecho tributario ante el reto de asignar valor fiscal a los servicios y transacciones económicas que derivan de la explotación de datos. In: Navarro, A. (dir.) Retos del Derecho Financiero y Tributario ante los desafíos de la economía digital y la inteligencia artificial, pp. 375–401. Tirant lo Blanch, Valencia (2020)
6. Sánchez-Archidona, G.: El Derecho tributario en la búsqueda de soluciones para los retos que plantean la robótica y la inteligencia artificial en la sociedad. Revista de Derecho UNED **27**, 17–56 (2021)

The Robotisation of Tax Administration

Rita de la Feria[1](✉) and María Amparo Grau Ruiz[2,3](✉)

[1] Tax Law, University of Leeds, Leeds, UK
R.DeLaFeria@leeds.ac.uk

[2] Financial and Tax Law, Complutense University of Madrid, Madrid, Spain
grauruiz@ucm.es

[3] Transnational Taxation, Northwestern University, Evanston, USA

Abstract. Developments over the last decade in the use of AI in tax administration have been nothing short of outstanding. Not only are taxpayers increasingly making use of automated systems in tax compliance, but perhaps more importantly, tax enforcement is increasingly reliant on new technologies as compliance-enhancing and fraud-prevention tools. However, whilst the use of AI brings very significant advantages to both the efficiency and the equity of tax systems, it also carries important risks. This paper identifies the development of a new AI fallacy within the tax policy sphere, namely that of unconstrained success: that the use of AI in tax compliance and enforcement can compensate for the deficiencies of the tax law. This paper considers the rationale behind the development of this fallacy, in particular the political and institutional dynamics involved in the approval of new tax legislation. It concludes that, maximising the benefits of the use of AI in tax compliance and enforcement requires departure from this fallacy, and the recognition of the wider dynamics of the tax policy-administration symbiosis.

1 Introduction

As it is often the case, it all started with tax law. Whilst today AI is used in many areas of law enforcement [1], the earliest best-known example of this use was the Taxman [2], an early-days AI system applied to corporate tax law enforcement, and sometimes referred to as "the father of AI and law" [3]. The use of AI-based technologies in tax compliance and enforcement, therefore, dates back to the 1970s [4], in the initial stages of AI, when tax law was regarded as particularly suitable for exploring its possibilities. This initial perception continued for the next decades, and between the 1970s and the 1990s, many more AI solutions were developed, focussing specifically on tax law [4].

It is undeniable, however, that the latest decade has been a time of unprecedented change in the use of AI in tax administration [5]. Not only are taxpayers increasingly making use of automated systems in tax compliance [6–8]; but perhaps more importantly,

An earlier version of this paper was presented at a conference held at Coventry University; we are grateful to the organisers and participants for the comments received therein; we also thank Modestas Kaseliauskas for fruitful discussions on several of the issues covered. The authors gratefully acknowledge funding from Horizon 2020, INBOTS: Inclusive Robotics for a Better Society, project 780073.

M. A. Grau Ruiz (Ed.): INBOTS 2021, BIOSYSROB 30, pp. 115–123, 2022.
https://doi.org/10.1007/978-3-031-04305-5_19

tax administrations are also increasingly reliant on new technologies as compliance-enhancing tools [9]. A recent OECD survey indicated that, in the 59 countries surveyed, more than 90% of business taxpayers were filing their returns electronically; 50% of tax administrations used digital assistants such as "chatbots"; more than 80% were using risk management analytical tools; and close to 75% used cutting-edge big data techniques [10]. Such rapid technological growth –which, to a large extent, mirrors the general growth in the use of AI in the public sphere [11] – is, however, bound to raise concerns. Although, initially anxieties over the developments in robotics focussed primarily on their impact on tax revenues [12], and their impact on tax administrations had gone largely unnoticed, this has changed in recent years with increased attention given to their (unintended) consequences.

In what follows, we provide a critical assessment of the impact of AI and robotics in tax administration. In Sect. 2, we consider the extraordinary growth of AI in tax administrations, and their many advantages as compliance-enhancement tools. In Sect. 3, drawing insights from public law and regulatory theory, we discuss the risks of using AI in tax enforcement, both from a generally regulatory perspective, and from a tax law-specific perspective. We then identify a new AI fallacy, namely that of unconstrained success, and consider the wider political and institutional dynamics that lead to its development. We conclude, in Sect. 4, that: (*i*) whilst the use of AI in tax administration has (very) significant advantages, it carries significant risks and limitations, and in particular it cannot compensate for deficient legal design; and (*ii*) maximising the benefits of the use of AI in tax administration requires the recognition of symbiotic nature of tax policy and tax administration.

2 The Growth of AI in Tax Administration

Whilst AI systems have been used for tax administration purposes since the 1970s, until recently, their usage was relatively restricted to specific functions, in a limited number of countries. The generalised use of AI at a global scale arguably started in the 1980s, with the digitalisation of tax compliance, such as electronic invoices and e-tax returns, but it was only in the 2010s that the use of sophisticated AI by tax administrations spread [13].

The use of AI in tax administration is now, not only pervasive but also diverse. Whilst there are many different AI tools in use globally, some of the most significant – and controversial – can be divided into three main types, as follows: risk assessment tools, real-time technology, and compliance assistance technology. The first two types can be broadly characterised as negative incentives to compliance, or anti-fraud mechanisms, the third as positive incentives to tax compliance, or compliance-enhancing mechanisms.

Risk assessment analytical tools have spread throughout the world [14]. Some of these tools focus on identification of high-risk taxpayers, including through big data sourcing and profiling, such as the Italian's FALCO system, or the Dutch XENON robot [15]. Others are aimed at improving the effectiveness of tax audits, known as computer-assisted audit tools and techniques (CAATTs), which have been implemented by several countries, including Australia, Finland, Germany, Indonesia and the US [16]. Further proposals concerning the application of AI to tackle international tax avoidance [4], and the development of AI systems that draws on big data to detect behaviour that may be hard for human inspectors to spot [17], have also been put forward.

One of the most promising – but also more intrusive – AI developments is real-time technology: the electronic matching-up of invoices in a data-warehouse, in real-time, so as to identify fraud and/or prevent it from ever taking place [18]. The process of matching-up invoices was not new, and indeed had been in place in South Korea since 1976, but AI proved crucial in overcoming running administrative and compliance costs [19]. In 2010, Israel successfully implemented a new online system, under which all invoice matching would be done electronically at a massive data- warehouse, followed by Portugal and Russia [20], and more recently Slovenia [15].

Compliance assistance technology is another area of significant growth in AI usage. Robots capable of performing tax compliance tasks, such as Turbotax, have already been developing for some years [21], and are likely to increase in the coming years [22]. Whilst tax compliance AI is often provided by private entities, its usage has been often encouraged by some tax administrations, with governments increasingly rely on AI to help the public understand and apply the law [8]. More recently, new AI-based predictors of tax law cases have also started appearing on the market, and increasingly being used by both taxpayers and tax administrations alike [23].

This outstanding growth of AI in tax administration can be attributed to a range of factors. First, the financial crisis over a decade ago sparked a period of renewed focus on tackling tax fraud, and non-compliance more generally. When faced with public finance pressures – namely the need to increase revenues, and minimise expenditure – many countries felt that improving tax enforcement and compliance was a politically easier – and importantly, also fairer – way of increasing revenues [25].

Second, human resources are expensive, and tax administrations are often underfunded [26, 27]. Pressure to improve tax compliance rates, is often coupled with drive to apply a cost-benefit analysis to tax administration, and an evaluation of enforcement elasticities: which administrative actions would result in the most revenue, at the lowest possible cost [28]. This approach to tax enforcement had some – perhaps unforeseen – effects, namely the development of a selective approach to tax enforcement, solely based on a revenue maximisation criterion [25]; but it also gave a strong impetus to AI implementation. An effective tool to increase revenues, with (comparatively) small administrative costs. Indeed, the probability of job automation for tax auditors and revenue agents has been estimated at 93% [29]

Third, AI systems have proved extremely effective as compliance enhancing tools [30, 31]. Not only are they effective in combating fraud – including newer practices themselves enabled by the digital economy [32] – but they are also effective in decreasing other forms of non-compliance, such as negligence or error [33]. Various new studies applying behavioural science to tax enforcement have shown how a significant part of non-compliance is non-intentional, so that removal of compliance frictions or small compliance nudges can have a significant impact [34–36]. It is therefore unspringing that, by facilitating compliance, AI can have a very significant impact on enforcement.

Finally, there is now also evidence of non-negligible spillover effects. In particular, well-designed AI can decrease two of the key downsides of tax administrative discretion [37], namely susceptibility to human cognitive biases and to noise, and opportunities for corruption. There is now consistent evidence to indicate that both cognitive biases and noise – the unwanted variability in judgments – are pervasive in administrative adjudication [38]. AI eliminates noise, which is a source of unequal treatment, and can counteract biases, if carefully designed. There is also evidence that AI decreases the scope for corruption [33, 39].

Overall, the use of AI in tax administration is likely to be a key positive factor in revenue mobilisation going forward –to some extent perhaps even offsetting the potential impact of automation on labour displacement [12]. Whilst the advantages are overwhelming, however, it is critical to be mindful not only of the potential risks of the use of AI by tax administrations, but also its limitations.

3 The Risks of Using AI in Tax Administration

There is now a growing literature on the ethics and regulation of AI, particularly in the context of its use by government agencies [40]. New technologies and innovation create, by nature, regulatory challenges, particularly in the context of traditional and reactive regulatory frameworks [41, 42]. Regulating AI is particularly challenging, however, not just due to the speed of the changes at stake, but because of their pervasiveness and the unforeseeability of future AI applications [43, 44]. It is unsurprising, therefore, that significant concerns have been raised as regards the use of AI. Some of these concerns apply generally to the use of AI both by private and government entities, others concern specifically the use of AI in tax administration.

General concerns as regards AI can be broadly divided into three types. First, there are fears vis-a-vis the impact of AI on discrimination and inequality. Whilst AI can correct human biases and noise if well-designed [45, 46], there is now strong evidence that many algorithms not only entrench the biases of its (human) designers, but augment them [47]. AI is often trained to identify correlations between characteristics and outcomes, using those correlations to predict future outcomes [49]. The problem is that correlation is not causation, and inferring causation from mere correlation can often lead to discrimination of specific groups, such as women or racial minorities [50, 51]. In tax administration, risk assessment tools are particularly susceptible to these profiling problems [52], as the recent Dutch scandal demonstrates [53]. Over the last decade, over 26,000 Dutch families were wrongly accused of fraud, after being singled-out by AI designed to detect large-scale fraud; more than half had an immigrant or vulnerable background. The Government resigned following a parliamentary report on the matter, but tax authorities remained largely unaccountable [54].

Second, there are also significant concerns regarding privacy rights, the management of big data, and the lack of legal remedies against AI. Some AI tools are extremely intrusive. They collect information from a variety of courses, including internet monitoring and communications monitoring, raising concerns over the respect for the right to privacy (Article 7 of the European Charter of Fundamental Rights); and based on that data, they predict future behaviour, arguably in violation of the right to be forgotten (Article 17 of the General Data Protection Regulation) [55–57]. Lack of due process and appeal of decisions or predictions made by AI is considerably worsened by humans' tendency to rely more on those decisions than on decisions made by humans, and thus a higher reluctance to challenge them [58]. In tax administration, real-time technology raises particular concerns in this regard. AI monitoring all transactions in real-time, has information on consumption habits of every individual in a given country, at any given time. From this information much can be inferred about the private life of those individuals, which is problematic not only in the context of autocratic regimes, but also of countries were cyber security is less strong and can be hacked by organised crime groups.

Finally, there are concerns about what is known as the ‘digital divide’, namely the consequences of increased robotisation of life on vulnerable groups, such as the elderly, who may lack the necessary skills or means to keep up with technological developments. For most – young, higher-income, higher-educated, tech-savvy, individuals – AI can be a convenient alternative to bureaucracy; but for the less tech-savvy elderly, or for those who lack the income to access digital services, or the language skills to understand them, the robotisation of life can have dehumanising effects [54, 59]. Tax compliance technology is particularly susceptible to these risks, and there is already evidence of divides emerging in countries, like the US, where compliance AI has been used the most [60].In addition to the above – now well-documented – risks, it is also important to consider the limitations of AI, what AI can and cannot do. This is something that, as opposed to the regulation and ethical consideration of AI, has received little or no attention in legal circles. The outstanding developments in AI use in the last decade has resulted in the subtle development of a new techno-fallacy [61], namely that of unconstrained success, or AI as the magic bullet. In taxation, this fallacy manifests itself in the growing belief that implementation of AI by tax administration can, on its own, solve the problem of tax non-compliance, and that tax administration can somehow compensate for the limitations of tax policy and legislation. Evidence of this approach is now widespread. In the US, the use of automated legal guidance by tax authorities has been designated as the “simplexity” approach: complex law presented as though it is simple, without actually engaging in simplification of the underlying law [62, 63]; in African countries, the drive to combat tax fraud has concentrated primarily on implementation of administrative measures that rely to a large extent on AI, with limited or no law reform [64]. Yet, as the African example demonstrates, although AI can bring significant advantages, it cannot *de facto* compensate for a deficient tax law.

How did this fallacy develop? Certainly, it can be partly explained by significant advantages of AI usage, and its proven track-record on compliance enhancement. Not all can be explained by its undeniable advantages, however – part of the explanation lies with the very real political dynamics of tax reforms. In essence, tax law reform is hard [65].

Approving new tax law is dependent upon lengthy legal procedures, often requiring qualified majorities that are premised on broad political consensus. Yet, consensus is often difficult to achieve. Not only do tax law reforms tend to have high political salience [66] – voters care about the outcome – but every reform will have losers, who, as opposed to winners, are usually politically loud and easily identified [67]. The political economy of taxation means that, throughout history, many elections have been lost or won as a direct result of failed tax law reforms [68]. This, together with institutional biases, such as path dependency, will often create legal entrenchment – the long-term persistence of legislation that has been associated with ineffective and obsolete laws [69].

On the contrary, implementation of new tax AI does not usually require lengthy legislative procedures – or often, even new legislation, thus often bypassing the need for legal reform. Perhaps surprisingly, the introduction of even the most intrusive of AI tools, such as real-time technology, has so far had very limited political salience, and even more limited public opposition. Very few people even know about it, let alone care about its risks; and, of course, it has many advantages. It is therefore easy to convince yourself that, actually, AI will resolve all your problems, and you do not really need tax law reform any way [70]. Coping with disappointment and other negative emotions is, after all, one of the most common causes of motivated reasoning [71]. Except, of course, you do actually need it.

4 Conclusion

The robotics revolution has firmly arrived to tax administration. The advantages are undeniable: if done right, AI can increase both the efficiency and equity of our tax systems; it is not, however, without risks or limitations. Whilst its impact on fundamental rights, ethics and regulation has been widely acknowledged, its effects on legal entrenchment, and maintenance of ineffective laws, have not so far been considered. This paper identifies a new AI fallacy, namely that of unconstrained success, and provides a rationale for its development. Whilst the particular difficulties in tax law reform present fertile ground for its development, it is unlikely that the fallacy is restricted to this area of the legal system. Rather, the likelihood is that it is now present, to a smaller or larger extent, in multiple areas of law – the higher the existing levels of legal entrenchment, the most likely it is that the implementation of AI will entrench the law further.

In taxation, regardless of the difficulties in legal reform, it would be unwise to put all our reform eggs in the AI basket. Ultimately, maximising the benefits of the use of AI in tax administration requires recognition of the wider dynamics of the tax policy-administration symbiosis [64]. Although they can bring many advantages, compliance-enhancing measures, such as new AI tools, are at their most effective when adopted in conjunction with a situational crime prevention approach to legal design: a legal system that has been designed to minimise tax non-compliance [72]. Tempting as it is to think otherwise, tax law reform remains a critical element of increasing efficiency and fairness in our tax systems – and not even robots can save you.

References

1. Freeman Engstrom, D., et al.: Government by Algorithm: Artificial Intelligence in Federal Administrative Agencies. Report Submitted to the Administrative Conference of the United States, February 2020
2. McCarthy, L.: Reflections on "taxman": an experiment in artificial intelligence and legal reasoning'. Harvard Law Rev. **90**(5), 837–893 (1977)
3. Susskind, R.: Pragmatism and purism in artificial intelligence. AI Soc. **3**(1), 29 (1989)
4. Kuzniacki, B.: The artificial intelligence tax treaty assistant: decoding the principal purpose test. Bull. Int. Taxation **72**(9) (2018)

5. Grau Ruiz, M.A.: Fiscal transformations due to AI and robotization: where do recent changes in tax administrations, procedures and legal systems lead us? Northwestern J. Technol. Intellect. Property **19** (2021, forthcoming)
6. Morse, S.: When will a tax compliance robot follow the law? Ohio State Technol. J. **16**, 278 (2019)
7. Mock, R., Shurtz, N.: The turbo tax defense. Florida Tax Rev. **15**, 443 (2014)
8. Blank, J.D., Osofsky, L.: Automated legal guidance. Cornell Law Rev. **106** (2021)
9. OECD: Tax Administration 2019: Comparative Information on OECD and Other Advanced and Emerging Economies, OECD Publishing, Paris (2019)
10. OECD: Tax Administrations Continue to Accelerate Their Digital Transformation (2021). https://www.oecd.org/tax/forum-on-tax-administration/news/tax-administrations-continue-to-accelerate-their-digital-transformation.htm
11. Calo, R., Citron, D.K.: The automated administrative state: a crisis of legitimacy. Emory Law J. **70**, 797 (2021)
12. de la Feria, R., Grau Ruiz, A.: Taxing robots. In: Grau, A. (ed.) Interactive Robotics: Legal, Ethical, Social and Economic Aspects, pp. 93–99. Springer, Heidelberg (2022). Ch 17
13. OECD: Unlocking the Digital Economy – A Guide to Implementing Application Programming Interfaces in Government. OECD Publishing, Paris (2019)
14. OECD: Tax Compliance by Design: Achieving Improved Compliance by Adopting a System Perspective. OECD Publishing (2014)
15. Ehrke-Rabel, T.: Big data in tax collection and enforcement. In: Haslehner, W., et al. (eds.) Tax and the Digital Economy, pp. 283–334. Kluwer (2019)
16. Darono, A., Ardianto, D.: The use of CAATTs in tax audits – lessons from some international practices. eJournal Tax Res. **14**(2), 506–526 (2016)
17. Chen, S.: China prepares to unleash artificial intelligence to catch tax cheats. South China Morning Post, 14 December 2019
18. Ainsworth, R.T.: Refund fraud? Real-time solution! digital security borrowed from the VAT, Brazil, Quebec & Belgium. Boston University School of Law Working Paper 12–15. de la Feria, R., Swistak, A.: A technology has the potential for significant applications in taxation, beyond tax administration, Designing a Progressive VAT", 2022, mimeo (2012)
19. Krever, R.: Combating VAT fraud: lessons from Korea. British Tax Rev. **3**, 329–341 (2014)
20. Giles, C.: Russia's role in producing the taxman of the future. Financial Times, 29 July 2019
21. Mock, R., Shurtz, N.: The turbo tax defence'. Florida Tax Rev. **15**, 443 (2014). The Turbo Tax Defence'
22. Morse, S.: Government-to-robot enforcement. University of Illinois Law Review, 1497–1526 (2019)
23. Alarie, B., Niblett, A., Yoon, A.: Using machine learning to predict outcomes in tax law. SSRN E-Journals (2017)
24. Alarie, B., Niblett, A., Yoon, A.: How artificial intelligence will affect the practice of law. SSRN E-Journals (2017)
25. de la Feria, R.: Tax fraud and selective law enforcement. J. Law Soc. **47**(2), 193–359 (2020)
26. Lederman, L.: The IRS, politics, and income inequality. Tax Notes **150**(11) (2016)
27. Gamage, D.: Creating Opportunity Through a Fairer Tax System: The Case for Taxing Extreme Wealth Holdings and "Real" (Book) Corporate Profits and for Improving IRS Funding". Indiana Legal Studies Research Paper 446 (2021)
28. Keen, M., Slemrod, J.: Optimal tax administration. In: IMF Working Paper WP/17/8 (2017)
29. Frey, C.B.: Osborne, M.A.: The future of employment: how susceptible are jobs to computerisation? Forecast. Soc. Change **114**(C), 254–280 (2017)
30. Grau Ruiz, M.A.: Fiscal transformations due to AI and robotization: where do recent changes in tax administrations, procedures and legal systems lead us? Northwestern Journal of Technology and Intellectual Property, p. 19 (2021, forthcoming)

31. Bird, R., Zolt, E.: Technology and taxation in developing countries: from hand to mouse. Natl. Tax J. **61**, 791–821 (2008)
32. Scarcella, L.: Tax compliance and privacy rights in profiling and automated decision making. Internet Policy Rev. **8**(4), 1–19 (2019)
33. Okunogbe, O., Pouliquen, V.: Technology, taxation, and corruption: evidence from the introduction of electronic tax filing. Am. Econ. J.: Econ. Pol. **14**(1), 341–372 (2022)
34. de Neve, J.E., et al.: How to improve tax compliance? Evidence from population-wide experiments in Belgium. J. Polit. Econ. **129**(5) (2021)
35. Hallsworth, M., et al.: The behavioralist as tax collector: using natural field experiments to enhance tax compliance. J. Public Econ. **148**, 14–31 (2017)
36. Halpern, D.: Inside the Nudge Unit – How Small Changes can make a Big Difference, pp. 112–115. WH Allen (2015). Although there is some evidence that people prefer moral discretion AI
37. Jauernig, J., Uhl, M., Walkowi, G.: People prefer moral discretion to algorithms: algorithm aversion beyond intransparency. Philos. Technol. **35**(2) (2022)
38. Sunstein, C.: Governing by algorithm? No noise and potentially less bias. Harvard Public Law Working Paper, pp. 21–35 (2021)
39. Sequeira, S., Djankov, S.: Corruption and firm behavior: evidence from African ports. J. Int. Econ. **94**(2), 277–294 (2014)
40. Ranchordas, S.: Experimental regulations for AI: sandboxes for morals and mores. Morals and Machines 1(1/2), and literature cited therein (2021)
41. Awrey, D.: Complexity, innovation, and the regulation of modern financial markets. Harvard Bus. Law Rev. **2**(2), 235–294 (2012)
42. Bennet Moses, L.: How to think about law, regulation and technology – problems with 'technology' as a regulatory target. Law, Innov. Technol. **5**(1), 1–20 (2013)
43. Ranchordas, S.: Experimental regulations for AI: sandboxes for morals and mores. Morals Mach. **1**(1/2), 86–100 (2021)
44. Smuha, N.: From a 'race to AI' to a 'race to AI regulation': regulatory competition for artificial intelligence. Law, Innov. Technol. **13**(1), 57–84 (2021)
45. Sunstein, C.: Governing by algorithm? No noise and potentially less bias. In: Harvard Public Law Working Paper, pp. 21–35 (2021)
46. Hacker, P.: Teaching fairness to artificial intelligence: existing and novel strategies against algorithmic discrimination under EU law. Common Market Law Rev. **55**(4), 1143–1185 (2018)
47. Mayson, S.: Bias in, bias out'. Yale Law J. **128**, 2218 (2019)
48. Kelly-Lyth, A.: Challenging biased hiring algorithms. Oxford J. Leg. Stud., 1–30 (2021)
49. Kleinberg, J., et al.: Discrimination in the age of algorithms. J. Legal Anal. **10**, 113–174 (2018)
50. Criado-Perez, C.: Invisible Women: Data Bias in a World Designed for Men. Abrams Press, New York (2019)
51. Rhue, L.: Racial influence on automated perceptions of emotions (2018). mimeo
52. Scarcella, L.: Tax compliance and privacy rights in profiling and automated decision making. Internet Policy Rev. 8(4) (2019)
53. van den Berg, S.: 'Dutch government quits over 'colossal stain' of tax subsidy scandal. Reuters, 15 January 2021
54. Ranchordas, S., Scarcella, L.: Automated government for vulnerable citizens: intermediating rights. University of Groningen Faculty of Law Research Report 11/2021 (2021)
55. Regulation (EU) 2016/679 of the European Parliament and of the Council of 27 April 2016, OJ L 119, 4.5.2016, pp. 1–88 (2016)
56. Ehrke-Rabel, T.: Bid data in tax collection and enforcement. In: Haslehner, W. et al. (eds.) Tax and the Digital Economy, pp. 283–334. Kluwer (2019)

57. Cardiell, L.: A robot is watching you: humanoid robots and the different impacts on privacy. Masaryk Univ. J. Law Technol. **15**(2), 247–278 (2021)
58. Citron, D.K.: Technological due process. Washington Univ. Law Rev. **85**, 1250–1314 (2008)
59. Ranchordas, S.: Empathy in the digital administrative state. Duke Law J. **72** (2022)
60. Bevacqua, J., Renolds, V.: The digital divide and taxpayer rights – cautionary findings from the United States. eJournal of Tax Research **16**(3) (2018). The term is borrowed from
61. Marx, G.: Rocky bottoms and some information age techno-fallacy. J. Int. Polit. Sociol. **1**(1), 83–110 (2007)
62. Blank, J.D., Osofsky, L.: 'Automated legal guidance'. Cornell Law Review **106** (2021)
63. Blank, J.D., Osofsky, L.: Simplexity: plain language and the tax law. Emory Law J. **66**, 189 (2017)
64. de la Feria, R., Schoeman, A.: Addressing VAT fraud in developing countries: the tax policy-administration symbiosis. Intertax 47/11, 950–967 (2019)
65. de la Feria, R., Walpole, M.: The impact of public perceptions on general consumption taxes. British Tax Rev. **67**(5) 637–669 (2020). For further analysis of the political dynamics of tax reforms
66. Gamage, D., Shanske, D.: Three essays on tax salience: market salience and political salience. Tax Law Rev. **65**, 23 (2011)
67. Fernandes, R., Rodrik, D.: Resistance to reform: status quo bias in the presence of individual uncertainty. Am. Econ. Rev. **81**, 1146 (1991)
68. Keen, M., Slemrod, J.: Rebellion, Rascals and Revenue – Tax Follies and Wisdom through the Ages. Princeton University Press, Princeton (2021)
69. Ranchordas, S.: One foot in the door: evidence-based limits on the legislative mandate. Hukim—J. Legisl. 207 (2018)
70. Kunda, Z.: The case for motivated reasoning. Psychol. Bull. **108**(3), 480–498 (1990)
71. Tykocinski, O., Steinberg, N.: Coping with disappointing outcomes: retroactive pessimism and motivated inhibition of counterfactuals. J. Exp. Soc. Psychol. **41**(5), 551–558 (2005)
72. Alldridge, P.: Criminal Justice and Taxation. OUP (2017) at 34. For a situational crime prevention approach

The Digitalization of Relations Between Citizens and Tax Administrations

Mercedes Navarro(✉)

University of Murcia, Murcia, Spain
mdes@um.es

Abstract. The purpose of this note is to analyze the position of citizens in the digital transformation process led by the Spanish Tax Administration. The intensive use of technology during the successive confinements imposed by the health crisis of COVID-19 is presented as an opportunity to move towards an electronica administration that is more secure, reliable and respectful with the rights of citizens.

1 Introduction

The pandemic has accelerated the digitalization process of Spanish society an also of its administrations. But in which direction?

In the field of tax application, the Spanish Tax Agency, since the 1990s, has been the spearhead in the construction of electronic administration. The intense technification has made it possible to offer citizens a wide catalogue of digital services. The technological capacity of the Tax Administration can also be perceived in the use of artificial intelligence as an essential piece for the prevention and control of tax fraud.

The powerful Tax Administration has led the digital transformation of tax relations. The intensive use of electronic media during the health crisis has highlighted the imbalance between the different actors in the digital relationship. Public authorities should take advantage of this context to advance in the construction of a legal status for the "digital citizen".

2 The Leadership of the Electronic Tax Agency

2.1 A Pioneer in the Massive Use of Electronic Tools

Spain occupies a prominent position in the graphs that measure the digital competitiveness of EU countries. For example, the Digital Economy and Society Index (DESI) [1]

Research project "Equity crowdfunding, innovative SMEs and digitalization of commercial companies: advances and challenges in European and Spanish Law" (PID2019-10463RB-I00/AEI/10.13039/501100011033), funded by the State Agency Research (Ministry of Science and Innovation), in the 2019 Call for "R+D+i Projects" of the State Programs for the Generation of Knowledge and Scientific and Technological Strengthening of the R+D+i and R+D+i System Oriented to the Challenges of Society.

M. A. Grau Ruiz (Ed.): INBOTS 2021, BIOSYSROB 30, pp. 124–129, 2022.
https://doi.org/10.1007/978-3-031-04305-5_20

places this country in second place in terms of digital public services. This data, however, is not indicative of an advanced digital society, since it is said that half of the population lacks basic skills and 8% have never used the Internet.

This advanced position of our Public Administrations has its maximum exponent in the Tax Agency. In the field of tax application, different applications and computer tools that require the collaboration of taxpayers have been successfully experimented with. Consider, for example, the assistance program for filing and paying the Personal Income Tax, whose electronical completion involves more than 20 million taxpayers.

Compared to other sector of the Public Administration, the technological and organizational superiority of the Tax Agency has made it possible to lead the transition from a relational model based on the traditional administrative procedure to a model in which the procedure, by default, is electronic. And the involvement of citizens and companies has been promoted with an increase in the obligations related to the electronic relationship with the tax administration.

2.2 From the Traditional Administrative Procedure to the Cooperative Relationship

Technology is also an essential component in advancing the transformation of the tax relationship towards a collaborative or cooperative model ("compliance by design").

This relational model, promoted by international bodies, takes center stage in the "Action Plan for a fair and simple taxation" [2], approved by the European Commission to face the solution to the health crisis.

The proposal of a legal relationship based on voluntary compliance is not new in our legal system, since the idea of a balanced relationship, with a greater role for taxpayers, was present in the gestation process of the General Tax Law (Law 58/2003) [3]. What is new is that it now runs in an electronic environment. And the Administration, in this context, has shifted the focus of attention from compliance with the tax obligation as an *ex lege* obligation towards obtaining and analyzing data ("big data").

The objective is to achieve a more exact knowledge of the taxpayers. To this end, the data pertaining to their taxable acts and their behaviour (past, present and future) are of interest.

The increase in the quantity and quality of data and his treatment through artificial intelligence components has completely changed the Administration's action schemes [4].

Data and technology are now the fundamental pillars to guarantee the correct application of taxes.

In general terms, the intensive use of technology by de Tax Agency makes it possible to identify three fully interrelated areas of action:

- Digital assistance services.
- Electronic processing.
- Data processing through artificial intelligence.

The realization of the right to information an assistance to taxpayers in the fulfilment of their obligations, which heads the catalogue of rights ad guarantees contained in art. 34

of the General Tax Law, has been favored by the numerous electronic services provided through the web portal (frequently asked question, databases, etc.).

In this area action, returning to the example of personal income tax, we have moved from the assistance program to generate the tax return (PADRE program) to the web form (Renta Web) [5].

The change that has taken place is not the simple substitution of a standardized model for another, now digital, that simply improves the user experience. The electronic form is a two-way data communication platform that transcends the very purpose of the declaration-settlement.

The form is no longer a help desk in the strict sense. The essential purpose of this electronic instrument is not to help the taxpayers to comply with his main tax obligation, but to generate data and optimize its use through algorithmic solutions.

The kind reading of the evolutionary process is that, for the taxpayers, the new generations of technology aspires to transform data into tax returns, streamlining tax management and reducing the intensity of subsequent control activity. But nevertheless, beyond providing a draft declaration to the citizen for easy confirmation, with digitization control is anticipated.

Taxpayer monitoring occurs at an early stage, as they provide data through the electronic office in compliance with their several formal duties (filing returns, keeping tax records, etc.).

The "Strategic Plan of the Tax Agency 2020–2023" [6] shows that "the line that separates assistance actions from preventive ones is becoming thinner". The technological reality allows prioritizing a strategy bases on prevention and control actions *ex ante* (sending letters, preventive visits, compliance, etc.).

This way of relationship between Administration and citizens is disturbing because, under the guise of cooperative compliance, proactive behaviour is imposed on citizens that goes beyond the duty of reasonable collaboration in a democratic State. And, on the other hand, this individualized monitoring escape the procedural architecture conceived by the General Tax Law of 2003.

2.3 A Potential Threat to Citizens' Rights

The digitalization of the tax administration is intended to facilitate the exercise of rights and compliance with the obligations by the citizens. Despite its undoubted advantages, its use by the Administration might turn into a possible threat to the rights and guarantees of taxpayers.

Different factors must be considered:

- An insufficient regulatory framework, which leaves a wide margin for action to the Administration.
- A multifaceted technological reality (assistance services, electronic processing and data processing).
- Imprecise legal limits, with a case law doctrine under construction.

In this context, it is not easy to determine how technology can affect the Administration's decision-making [7]. This opacity can lead to an imbalance in the tax relationship, and maybe leave citizens without the ability to defend themselves or challenge claims against them, a situation intolerable in an advanced democratic society.

3 Limits to Technology

3.1 General Principles and Rights

The technological environment is subject to the limitations established by the Constitution and the laws.

Such assertion is, however, insufficient to provide an adequate response to the challenges posed by the digitalization of relations between citizens and the Tax Administration.

The change in the relational model promoted by the Tax Agency is not adequately regulated to avoid uncertainty and safeguard the rights and guarantees of taxpayers. However, some case law pronouncements are beginning to shape the technological reality emphasizing its instrumental function, subject to the principles and rights enshrined in the Constitution [8].

The horizon that opens after the digital immersion produced by the pandemic makes it necessary to review the legal status of taxpayers to adapt the list of classic rights and guarantees to the digital ecosystem.

3.2 Take off of Digital Rights

Computing technology is present in the Spanish Constitution, where it supports a new generation of rights, the so-called "digital rights". This legal aspiration takes shape in Organic Law 3/2018, of December 5, on the Protection of Personal Data and guarantee of digital rights.

This regulation does not intend to establish a complete and well-defined catalogue of the rights of citizens in the digital environment, but it does make visible the Spanish legislator's commitment to strengthening the legal status of the digital citizen.

Although the legal text does not address the risks that technology entails for the rights of citizens in their electronic relations with Public Administrations, there is no doubt that its postulates are projected to this context, since compliance with the rights and freedoms applicable to The Internet is a must for all digital service providers.

The Charter of Digital Rights [9], presented by the Government within the framework of the Digital Spain Plan 2025 [10], expressly formulates the rights for citizens who are electronically related to the Administrations.

Recognition of the principles of technological neutrality and general access to the Internet is essential to avoid discrimination for technological reasons. More complicated will be the creation of a possible right of access to the algorithm, the prohibition of algorithmic discrimination, the right to be forgotten in the use of data by the Administration, or the right to human supervision in automated actions. Neither is it easy to implement the culture of proactivity in accordance with the standards required by the regulations on the protection of personal data.

The catalogue of digital rights is yet to be defined, but the task is urgent and the formulation of these rights in the Digital Rights Charter is, clearly, a significant advance in the digital transformation of our society.

Recognition of the principles of technological neutrality and general access to the Internet is essential to avoid discrimination for technological reasons. More complicated will be the creation of a possible right of access to the algorithm, the prohibition of algorithmic discrimination, the right to be forgotten in the use of data by the Administration, or the right to human supervision in automated actions. Neither is it easy to implement the culture of proactivity in accordance with the standards required by the regulations on the protection of personal data.

The catalogue of digital rights is yet to be defined, but the task is urgent and the formulation of these rights in the Digital Rights Charter is, clearly, a significant advance in the digital transformation of our society.

4 Conclusion

There are many open questions in the field of technology applied to taxes, which will have to be answered so that the transition from the traditional relational model to a digitalized one does not entail a deterioration of the rights and guarantees of taxpayers.

It is necessary for the powerful "electronic Tax Agency" to evolve towards a more transparent, participatory and reliable digital tax administration.

Acknowledgment. This note is part of the research project "Equity crowdfunding, innovative SMEs and digitalization of commercial companies: advances and challenges in European and Spanish Law" (PID2019-10463RB- I00/AEI/10.13039/501100011033), funded by the State Agency Research (Ministry of Science and Innovation), in the 2019 Call for "R+D+i Projects" of the State Programs for the Generation of Knowledge and Scientific and Technological Strengthening of the R+D+i and R+D+i System Oriented to the Challenges of Society.

References

1. https://administracionelectronica.gob.es/pae_Home/pae_OBSAE/Posicionamiento-Internacional/Comision_Europea_OBSAE/Indice-de-Economia-y-Sociedad-Digital-DESI-.html
2. https://eur-lex.europa.eu/legal-content/EN-ES/TXT/?from=ES&uri=CELEX%3A52020DC0312
3. https://www.boe.es/buscar/act.php?id=BOE-A-2003-23186
4. Grau Ruiz, A.: Riesgos y oportunidades en la creciente digitalización fiscal. Revista Técnica Tributaria (130) (2020)
5. https://sede.agenciatributaria.gob.es/Sede/Renta.html
6. https://sede.agenciatributaria.gob.es/Sede/planificacion/plan-estrategico-agencia-tributaria-2020-2023.html
7. Grau Ruiz, A.: La supervisión de los procesos de automatización de la administración de los tributos: prácticas en España y en Estados Unidos. Revista Técnica Tributaria (135) (2021)
8. Navarro Egea, M.: Hacia un entorno digital más garantista: las relaciones tributarias electrónicas, Aranzadi, p. 65 et seg (2021)

9. https://www.lamoncloa.gob.es/presidente/actividades/Documents/2021/140721-Carta_Derechos_Digitales_RedEs.pdf
10. https://portal.mineco.gob.es/ca-es/ministerio/estrategias/Pagines/00_Espana_Digital_2025.aspx

Check for updates

Robotization in Excise Duties: Has It Arrived with the Immediate Supply of Accounting Books (SILICIE)?

Antonio Fernández de Buján y Arranz[1,2](✉)

[1] KPMG Abogados, Madrid, Spain
a.fernandezdebujan@ucm.es
[2] Complutense University of Madrid, Madrid, Spain

Abstract. The new system for the immediate supply of information for excise duties, framed in the process of progressive implementation of robotics and artificial intelligence in the field of Tax Law, which which can no longer be turned back, has meant that companies obliged by the SILICIE have had an important job of adapting and parameterizing their accounting to the new formal and material requirements of the Spanish Tax Administration Agency (hereinafter "AEAT"), in a reduced period of time, as well as internalizing it.

1 Introduction

The progressive adaptation of the tax system to the challenges currently posed by the implementation and expansion of robotics in different areas in the European Union is an unavoidable necessity. Robotics is a transformative technology of the economy and society. At this time, many States are discussing how to combine social and fiscal policy so that the risks in the distribution of costs and benefits derived from policies to stimulate innovation are distributed in a manner consistent with the objectives of human sustainability [1].

The fourth Industrial Revolution comes hand in hand with robotization and artificial intelligence as a fundamental element of this transformation. For the Tax Administration, this revolution is connected with the growing volume of data (Big Data), the use of algorithms to process them, and the massive interconnection of systems and devices [2].

In Spain, in the area of Excise (hereinafter "IIEE"), starting from January 1, 2020, the taxpayers, who are the owners of factories, tax warehouses, reception warehouses, and vinegar factories, must supply through an accounting system in computer support through the electronic headquarters of the AEAT [3], the processes, movements and stocks of the products subject to the IIEE and, where appropriate, of the raw materials necessary to obtain them, including differences that become apparent during storage, manufacture or circulation [4].

2 Material and Methods

It should be remembered that the aforementioned supply is not in itself novel since it was an obligation contained in article 50 of the Excise Tax Regulation. Now, what

M. A. Grau Ruiz (Ed.): INBOTS 2021, BIOSYSROB 30, pp. 130–134, 2022.
https://doi.org/10.1007/978-3-031-04305-5_21

really represents an important differentiating element in relation to the already existing obligation, and now framed, in my opinion, in the strategic use of data and artificial intelligence in the public sector, is the considerable shortening of delivery times, in addition to the significant administrative burden that has been placed on taxpayers in order to adapt their computer systems to the new requirements of the AEAT. That is, with the SILICIE, more information should be provided, in order to <control and prevention of tax fraud>>, and in less time.

With regard to robotics [5] and taxation, we are faced with an unquestionable reality that, entails the appearance of an added problem that, although it is not new, is still worrying: the so-called, in Keynes's words , technological unemployment, that is, the destruction of a significant number of jobs –especially those that, due to their characteristics, have lower qualifications–, a circumstance that, as has been well pointed out, generates, in turn, a additional concern, which is the loss of possible economic resources (particularly, in the form of social security contributions, as well as in the field of personal income tax) that will make it difficult, not only in the future, but also in the present, both objectives of fiscal consolidation, such as, and above all, the adequate maintenance of the Welfare State [6].

However, in my opinion, with regard to the tax field, the interpretation of the data will always be necessary, and human reasoning, which logically goes beyond algorithms and mathematical formulas and, to this day, cannot be attributed robots.

Compliance with this obligation to keep the accounts in the case of factories, bonded warehouses, reception warehouses, tax warehouses or vinegar factories will be carried out by means of an accounting system in computer support, through the AEAT electronic headquarters, with the electronic supply of the accounting entries that will be made through a web service or, where appropriate, an electronic form.

In relation to the supporting documentation and information related to the operations and movements carried out, art. 10 of the SILICIE Order establishes, that the owners of factories, tax warehouses, reception warehouses, and vinegar factories must always have available to the inspection and intervention services of the Tax Administration, the following documentation and information:

a) Justifying documentation and information related to the manufacturing and/or transformation operations carried, detailing the start and end dates, the raw materials and products used, the products obtained and the number of the work or operation manufacturing or transformation report.
b) Regulatory and commercial documentation justifying the movements, processes, and stocks, as well as details of the inventories of the products subject to the Excise Taxes and the raw materials necessary to obtain them.
c) Justifying documentation and individualized information of each of the operations that have been supplied in a single accounting entry in aggregate form.
d) The accounting system in computerized support from which the electronic supply of the accounting entries has been made.

On this particular, section 2 of art. 10 of the SILICIE Order adds that the supporting documentation and information related to the movements, products and stocks, processes and operations carried out, as well as, where appropriate, the accounting system, must be available to the inspection services and, where appropriate, case, of intervention, in the establishment itself while the limitation period of the Administration's right to determine the tax debt through the appropriate liquidation has not expired.

In the case of establishments that extend over the territorial scope of more than one management office, the indicated documentation will be kept in the place where the management and control centre of said establishment were located.

For these purposes, the required documentation may be made available in computerized support provided that it can be consulted without delay from the establishment itself.

3 Results

As regulated in art. 50 RIIEE, specifically in the wording of the article in force until December 31, 2019, the accounting books by means of computer support must be submitted to the management office, electronically in the following cases and deadlines:

a) In the case of factories, tax warehouses and reception warehouses, within the month following the end of each settlement period.
b) In other cases, for example, in the case of pharmaceutical companies that used alcohol in their production process, within the month following the end of each quarter.

In SILICIE it could be said that the system already created for VAT purposes, Immediate Supply of VAT Information, has been reproduced, *mutatis mutandis*, with the other indirect taxes par excellence, but only in the area of excises manufacturing taxes, within which are included the excise tax on hydrocarbons, alcohol and tobacco, respectively.

For these purposes, the AEAT introduced in the period of amendments to the Draft Order a particularity for the supplies for the month of August which, in my opinion, is very appropriate. In this way, the owners of establishments that, having a computerized accounting system, have opted for the keeping of the accounts and the electronic supply of the accounting entries from the same, may consider the month of August as non-working exclusively for purposes of the aforementioned supply obligation.

4 Discussion

In general terms registration obligations are not changed following the SILICIE. However, this system does substantially affect supply obligations that previously occurred within the month following the end of each settlement period and with the new system is five business days.

For these purposes, the supply period will be computed from the date the entry was recorded in the establishment's accounting system. In the event that the accounting entry had not been registered, in general, within the twenty-four business hours following the moment of the movement, operation or process that is the subject of the entry, the supply period will be calculated from the end of said period.

Note that these terms, in general, are reduced in the event that the owners of the establishments obliged to SILICIE do not have an accounting system in computer support in which they record the movements, processes and stocks of the products subject to the Excise duties and the first matters necessary to obtain them or dispose of the same have not opted for the keeping of the accounting and the electronic supply of the accounting entries from the same.

For this case, that is, if no prior accounting takes place, the delivery times are those established in art. 5. In this case, the electronic supply of the accounting entries through the electronic Headquarters of the State Tax Administration Agency must be carried out within twenty-four business hours following the moment of the movement, operation or process that is the subject of the entry.

Some operators that meet certain requirements, among others, refineries, pharmaceutical companies, breweries, intermediate products, wine cellars and fermented beverages and tax warehouses constituted by a network of natural gas pipelines, will be applicable deadlines and accounting and/or supply obligations that differ from the general regime.

5 Conclusions

By way of conclusion, it should be noted that this new system for the immediate supply of information for excise duties, framed in the process of progressive implementation of robotics and artificial intelligence in the field of Tax Law, which now has no turning back, has meant that the companies bound by it have had an important job of adapting and parameterizing their accounting to the new formal and material requirements of the AEAT, in a short period of time, as well as internalizing it.

As has already been indicated, the information contained in the SILICIE was already provided electronically by the taxpayers, but within a more reasonable period of time.

For sure, the entry of artificial intelligence algorithms to the AEAT and, specifically in the field of IIEE, as a consequence of a non-static view of the data obtained by the AEAT, have come to stay. In sum, we can only predict a satisfactory development both for the Tax Administration and for those obliged to keep SILICIE [7].

In this sense, taxation has been affected by this phenomenon in many of its facets, because traditional tax models are being replaced by new principles and rules of subjection to taxation and also by obtaining and analyzing data through new IT tools and artificial intelligence techniques that tax administrations benefit from.

Acknowledgement. This article is part of the following Research Projects: <<Trascendencia jurídico-financiera de las auditorías de sostenibilidad mediante la gestión inteligente de datos>> (ref. PID2019-105959RB-I00), within the framework of the National I + D+ I Plan of the Ministry of Science and Innovation, whose main researcher is Prof. Dra. Dª. Amparo Grau Ruiz.

<<Los derechos y garantías del contribuyente ante la Hacienda pública del siglo XXI>> funded by the Institute of Fiscal Studies and whose Principal Investigator is Prof. Dr. D. Juan Calvo Vérgez.

References

1. Grau Ruiz, M.A.: La adaptación de la fiscalidad ante los retos jurídicos, económicos, éticos y sociales planteados por la robótica. Nueva fiscalidad (4), 36 (2017)

2. Serrano Antón, F.: Fiscalidad y Robótica: funcionalidades disruptivas en el Derecho tributario. en VVAA. Fiscalidad e Inteligencia artificial: Administración tributaria y contribuyentes en la era digital" (Coord. SERRANO ANTÓN, F.), Aranzadi (2020). https://proview.thomsonreuters.com/
3. de Buján y Arranz, A.F.: El suministro de información de los impuestos especiales en el plató, Cinco Días, 16 de octubre de 2019. https://cincodias.elpais.com/cincodias/2019/10/15/legal/1571142247_444411.html
4. Calvo Vérgez, J.: La nueva obligación de suministro inmediato de Libros contables de Impuestos Especiales (SILICIE): principales rasgos configuradores. Quincena fiscal (9) (2020)
5. Falcón Pulido, A.: Aproximación a la tributación de los robots. Revista Técnica Tributaria (132), 239
6. Álvarez Martínez, J.: Robótica y fiscalidad: unas breves consideraciones. Quincena Fiscal, (3) (2021). https://insignis.aranzadidigital.es/
7. Grau Ruiz, M.A.: Digitalización, inteligencia artificial y economía ircular (dirs. García Novoa, C.; Vivel Búa, M), Aranzadi Thomson Ruters, pp. 51–63 (2021)

Artificial Intelligence and Robots: The Role of Tax Legislator, A Conundrum to Solve

M. G. Ortoleva(✉) and S. M. Messina

Tax Law, University of Verona, Verona, Italy
mariagrazia.ortoleva@univr.it

Abstract. The paper aims to investigate if and how the tax legislator can intervene in the implementation phase of artificial intelligence and robots so that their use in the production of goods and services generates a social utility for the community. The questions that arise are different. In particular, it is necessary not only to understand whether to encourage or tax the use of robots/AI but it is also necessary to explore options for the introduction of legal tools to improve both the relationship between citizens and tax administrations and the system's efficiency in monitoring and tackling evasion and fraud.

1 Introduction

The development of robotics and artificial intelligence profoundly changing production as well as service delivery models (with automated and interconnected productions) and calls for "rethinking" work organization "methodologies" in both the private sector and the public administration.

The ongoing technological revolution "affects" the several aspects of the tax system. In this context, it seems reasonable to undertake an inquiry into the tax legislator's role. Recent economic and social changes beg the crucial question of whether the time has come to consider the taxation of new forms of wealth produced by artificial intelligence and robots. At the same time, it is also essential to understand whether there are the conditions and guarantees for a complete process of "digitization" of the fiscal administration.

2 Taxing or Incentiving?

1. The ability to support an innovative process is widely regarded as a decisive factor in keeping up with the most advanced economies. And that's precisely the reason many countries, including Italy, have chosen to encourage, also through the provision of favourable tax treatments, investments in robotics and in the development of interconnected production systems using AI.

Accordingly, taxing robots or the individuals who benefit from them would constitute a radical change of direction in tax policies currently in force in this field. On the one hand, this idea is underpinned by the requirement to compensate what would

M. A. Grau Ruiz (Ed.): INBOTS 2021, BIOSYSROB 30, pp. 135–142, 2022.
https://doi.org/10.1007/978-3-031-04305-5_22

seem to be the most significant of the negative externalities resulting from the use of robots, namely, the so-called "technological unemployment", on the other, by the need to implement redistributive policies. It is believed that the robotic process automation with implemented artificial intelligence will lead to massive job losses, that the operating mechanisms of the market would not be able to cope with the consequent reduction in general welfare resulting from the ongoing revolution and that only the State may address and rectify this market failure, by implementing adequate taxation policy instruments. Moreover, recourse to such instruments would be required to address the decline in tax revenues due to the use of AI/robots. More specifically, to cope with a reduction in tax revenues collected by taxing regular employment income, but also to cope with a reduction in tax revenues (as pointed out by Chand, Kostic, Reis) resulting from VAT collection, in view of the fact that robots and various forms of AI, unlike humans, do not buy cars, clothes, food, etc.

In essence, the idea of taxing the use of robots that replace human workforce or even taxing the robots themselves has its origin in the need of a State intervention. This intervention would be aimed, on the one hand, at remedying the situation of growing income inequality, and on the other, at making up for the high technological unemployment by slowing down in the speed of replacement of individuals with the AI/robots (in this paper with this expression we refer to those technologies capable of performing, even without human intervention, routine tasks or activities more complex previously carried out only by humans being).

Starting from this assumption, various theories have been elaborated. Some (theses) are more traditional, like taxation of property namely providing a tax that would be levied on the ownership and could be based on a flat rate or vary according to the type of robot so-called "object tax". Others are more innovative, such as the introduction of a direct tax on robots (see Oberson 2019, at p. 132, which suggests considering in the nearest future robots as taxpayers per se and introducing a "robot Income or Revenue Tax (so-called robot Income or Revenue Tax). Finally, some authors enquire about different forms of robot use taxation and propose several designs. One option is to allocate an "imputed salary" to the use of robots that would be equivalent to the salary that an individual performing the same work would have received. Another is to provide for "automated" companies a less favourable treatment than the normal taxation regime. In this perspective, it has been proposed for example both a relevant increase in corporate income tax rates for automatized businesses (Abbott and Bogenschneider, at pp. 172–173) and the reduction of depreciation allowances for corporation tax purposes, as well as a partial denial of input VAT deductions for the acquisition of robots, depending on the degree of the firm's automation (Abott and Bogeschneider). Yet others hypothesize the introduction of new forms of taxation, this is the case of the "Robot tax" based on the created value to profits obtained by businesses considering a rise in their profits due to the implementation of new technologies (Lexer and Scarcella 2019).

Notwithstanding the highlighted structural differences, the proposed options, except for the direct tax on robots, employ traditional legal categories to tax a new and peculiar economic phenomenon. On closer inspection, many of these continue to be based on the notion of robots as tools in the hands of other actors, forgetting that need to tax arises in the degree of autonomy of the new robots.

2. Having said this, it is necessary to carry out some reflections from an economic and, above all, legal point of view. A first issue to be addressed concerns the compatibility of new forms of taxation of 4.0 enterprises with the principles underlying modern tax systems or with their constitutional justification, starting with the ability to pay principle.

As pointed out, from a purely economic point of view, the taxation of AI/robots should find its justification in the need of compensating negative externalities generated by entrepreneurial choices inherent in the use of AI and robots and in the awareness of the need for State intervention in the economy. All the proposals discussed above are based on the idea that employing AI/robot is causing, at least during the transition period, widespread unemployment and they aim at reducing the laying off and/or replacement of employees by AI/robots. However, it should be noted that there is no consensus regarding the implications of automation and, in particular, how many jobs would be replaced in the future nor how many would be created as a result of automation.

It will suffice to recall that, while some believe that the implementation of AI and robotics in production processes will result in a significant loss of employment (which, in turn, will lead to a consequent mass layoff/unemployment, (Frey and Osborne 2013), nevertheless, a report by Oxford Economics underlines that "it would be simplistic to characterise robotization as only a destroyer of jobs". In particular, the report, on the one hand, indicates that "jobs are both created and destroyed through the increased use of automation and industrial robots", on the other, underlines that "a faster adoption of robots has a positive impact on both short- and medium-term growth. "For example, boosting robot installations to 30% above the baseline forecast by 2030 would lead to an estimated 5.3% boost in global GDP that year". (Oxford Economics, June 2019. See also Autor 2015). In the view of Chand, Kostic and Reis "The one-off statistic might demonstrate that, after a technological revolution, society needs an "adjustment" period for its own reorganization and then it starts benefiting from the new technologies. In other words, employment levels increase after the "adjustment" period. Arguably, this conclusion may also indicate that employment levels could also increase during or after Industry 4.0 following an initial 'adjustment' period".

3. Obviously, such theories have evident repercussions from a juridical point of view.

Should one believe adequately and sufficiently demonstrated that the use of robots and new AI technologies will lead to a social detriment - that is, in case it was reasonable to assume that the use of AI and the robots will entail a "stable" loss (not a temporary loss) of jobs and that this unemployment has a causal link with the mere use of robots - then it could be advisable to introduce a sort of ablative tax levy and disregard the ability to pay principle. These are levies whose legal standing stems from the damage that the use of the robot produces or that it could produce. In essence, these would be levies with a "reparative" purpose (as claimed by some about environmental taxes). This hypothesis, however, should be abandoned both because of the uncertainty that exists regarding the consequences that the industrial and technological revolution is having on society (there is no irrefutable evidence of the existence of an immediate causal link between the use of robots and the loss of employment), and because it seems difficult not to recognise a co-responsibility of the State. The latter could be held liable for failure to introduce sufficient training and re-training worker incentives and appropriate policies (fiscal or otherwise) to avoid such negative externalities.

Disregarded this hypothesis, the introduction of a levy/tax on robots, can only be justified on the basis of the ability to pay.

According to doctrine, "revenue interest" alone cannot justify taxation. It must be considered incompatible with the ability to pay principle to tax a subject in the absence of specific parameters designed to represent an effective ability to pay, to pursue an interest allegedly to be preached as superior (since it belongs to the State).

Having said that, it is well known that the above-mentioned principle requires that only facts suitable for demonstrating a certain economic strength should be subject to taxation. According to the Italian Constitutional Court, the ability to pay "must, in principle, be derived from any index of wealth, according to the assessments of the legislator, except for the control of constitutionality in terms of arbitrariness and irrationality".

Based on that parameter, the lawful taxation of owners or subjects who use robots seems to be subordinated to whether robots/AI - unlike other "means of production" - can be considered as goods that represent an economic value in themselves, capable to justify an "additional" tax which falls on the subject who employs or handles them. A similar conclusion could perhaps apply to robots or so-called "smart" machines that "take decisions, interact with third parties independently "of external control or influence" (recital AA). We refer to those robots for which the European Parliament had suggested the idea (later rejected) of "applying electronic personality", on the assumption that "the more autonomous robots are, the less they can be considered to be simple tools in the hands of other actors" (European Parliament resolution of 16 February 2017 with recommendations to the Commission on Civil Law Rules on Robotics (2015/2103(INL)).

Finally, the assessments concerning the compatibility with the ability to pay principle relating to direct taxation of robots appear more complex since, in this case, the profile of the robot's tax subjectivity is also at issue.

In this regard, it should first of all be clarified that, according to the author, the choice of the EU not to recognize, at least for the time being, legal subjectivity under civil law to robots, does not prevent these from being qualified as taxable persons, such as for income tax. The author believes that also 'entities' not qualified as legal persons under different legal frameworks could be taxable persons, but in so far as these 'entities' enjoy asset autonomy. Indeed, tax subjectivity is conditional upon the asset autonomy of the qualifying entity. This is because only asset autonomy is a suitable indicator of the ability to fulfil future tax debts (so-called financial capacity).

Having said this, in practice, however, it seems impossible that robots can have an autonomous financial capacity as long as they do not have civil law recognition.

In any case, apart from the mentioned issues, robot taxation would remain difficult to reconcile also with a further element of the ability to pay principle. The latter is also the criterion based on which public expenditures are allocated to taxpayers. Accordingly, it postulates that the person bearing the tax enjoys the "benefits" of belonging to the 'community'.

Therefore, since it does not seem reasonable to equate them to an individual, the author submits that robots should not be deemed independent taxable persons (similar to human beings). This line of reasoning would also hold in the event civil law recognized the independent imputability of robots as if they were people (so-called electronic personality). Acknowledgement by civil law would not alone be sufficient justification for taxation.

Different conclusions could be reached in the event that, instead, should civil law awards to robots a legal personality similar to legal entities such as corporations. The taxation of legal persons - as known - is independent of their participation as members of the community. In this case, if we assume that the civil law choice to recognize robots as legal entities arises from the intent to make them liable for damages they could cause, then they should have patrimonial autonomy and the accessibility of the funds generated by the robots by the respective owner or user should be restricted. Such liability should assume a certain economic substance. The conditions would therefore exist to qualify robots as liable to a tax functionally equivalent to a corporate income tax.

4. However, the fact that taxation of robots can be justified based on the ability to pay principle does not mean that an implementation of robot taxation would also be "appropriate". Indeed, since this is a clear political choice, several factors must be taken into account, including those of an international and economic dimension.

In this respect, it is common ground that the introduction of specific forms of levies on AI or robots by individual States would negatively affect their own competitiveness and, most probably, would induce many companies to relocate to other jurisdictions. As has been pointed out "Globalization has reduced the fiscal space to many governments and has made the tax systems less progressive" (Tanzi 2015).

The difficulties involved in identifying the scope of any type of taxation in the first place and the possibly distortive effects of such taxation should, then, not be overlooked.

Furthermore, the findings of some studies which conclude that "high tax rates on all incomes and especially on incomes from capital sources have a negative impact on growth and employment" (Tanzi 2014), also argue against the introduction of new forms of taxation. If this is true, then an increase in taxation tout court on automated enterprises may not have a positive effect on employment rates.

Taking all these considerations into account, it seems inappropriate to interpret the choice on the role of States in terms of a stark contrast between encouraging or taxing automation, namely between financing undertakings that invest in new technologies or taxing them, to replace lower revenues from labour income and have "new" sources of revenues to support employment and/or redistribute wealth.

On closer inspection, the two objectives - promoting economic development and supporting employment - are only apparently opposed.

Automation is crucial to ensure the competitiveness of the entire "country system" and, therefore, to sustain employment. After all, a condition of widespread unemployment could negatively affect the target markets of undertakings, even those that are automated.

These are the reasons that suggest considering that, as things stand, the issue of the role of the State should first and foremost be posed in terms of an appropriate allocation of resources.

In this perspective, it seems essential for the State to adopt a proactive fiscal policy aimed not only at encouraging investments in AI but also at supporting employment through incentives to provide the new skills (that "automated" society will need) and through the reduction of the so-called tax wedge. The latter reduction, on the one hand, could be effective in supporting labour demand and, on the other hand, would also be an incentive for workers.

The analysis carried out so far leads, in conclusion, to agree with the EU's choice to reject a proposal to implement a tax on robots. In contrast, the decision of the Italian legislator to extend only until 2022 the incentive of tax credit (introduced in 2020), aimed at supporting investments by companies to train employees on issues relevant to the technological and digital transformation 4.0, would be open to criticism. All the more where it is considered that the resources allocated to incentives for automation have increased.

3 Digitalization of Fiscal Administration: Self-compliance Instruments and Means of Contrast to Illicit Tax Practices

The ongoing revolution has also drawn attention to the benefits and potentialities associated with the digitalization of public administrations, including tax administrations.

The European Committee, as it is well known, has already prepared a plan in 2016 to encourage the digitalization of the administrative authorities of the Member States ("EU eGovernment Action Plan 2016-2020 Accelerating the digital transformation of government" (COM/2016/0179)). Also, recommendations aimed at fostering the digitization of the fiscal administration have been issued by the OECD.

In fact, new technologies are likely to be used in two other fundamental directions.

The first is to encourage self-compliance by taxpayers and to reduce compliance costs and bureaucratic burdens. The second is to provide the fiscal administration with more effective instruments for the repression of tax offences.

As for the first issue, the use of new tools in the field of tax liability determination and settlement is already a reality not only for the performance of natural persons but also for those charged to companies. We refer here to the electronic invoicing system which was followed by settlements, pre-filled VAT accounts and preparation of the VAT return.

As for the second issue, on the one hand, the use of these technologies can contribute to the achievement of the objective of more correct and efficient functioning of the administrative action, on the other, it raises concerns on the effective "protection of personal data" and the respect of the principle of "proportionality". The collection of an impressive amount of data "stored" by the fiscal administration and the processing thereof may be "disproportionate" concerning to the need to minimize the acquisition, processing and storage of personal data, including those of the taxpayer.

It would therefore be necessary - as the case law of the Court of Justice of the European Union and that of the European Court of Human Rights also states - to ensure an adequate balance between the principle of good administration and the fiscal interest, on the one hand, and the taxpayer's right to privacy, on the other. To this end, it seems necessary to lay down the rules for the correct and secure acquisition of data and at the same time to provide immediate and effective protection for the taxpayer, to avoid possible abuses by the administration.

In particular, systems capable of collecting, processing or storing personal data must be designed from scratch in such a way as to ensure the protection of personal data (the principle of "data protection by design and by default"), the security, secrecy and inviolability of the data by third parties other than the main contracting authority. Moreover, it is mandatory to ensure that the data entered can be modified at any time. In this regard, it is to remember that the new tools, which are made available by technology, is based on the final and ultimately definitive unchangeability of the data entered (as is the case for "resilient" systems, such as data files formed based on distributed Ledger technology and in particular blockchain).

Issues of taxpayers' protection are even more urgent in relation to the instruments of preventive control based on AI available to the fiscal administration "without the agreement of the taxpayer".

If there is no doubt about the potential offered by predictive intelligence in the fight against the phenomena of evasion and fraud, it is at the same time essential that the legislator establishes the "available" tools, the conditions for using them and, last but not least, their evidential value. Regarding the latter, there is the risk to consider the results of predictive analyses of possible evasive behaviour as a presumption of evasion, while they should be used only as triggers for further substantive checks.

It is, therefore, necessary to provide the national systems with legislation on the protection of personal data specifically devoted to tax relationships, based upon a reasonable balance between, on the one hand, the principles of good administration and fiscal interest and, on the other, the taxpayer's right to privacy for which immediate and effective protection must be guaranteed. The intent is to make the amount of data that taxpayers are obliged to share with the fiscal administration in line with the principle of proportionality.

4 Conclusion

The changes generated by the development of robotics and artificial intelligence pose new challenges and, at the same time, represent an opportunity. Like any "industrial" revolution, even the one in place at the moment can generate both positive and negative externalities.

The tax legislator could play a fundamental role in determining the overall effects on the community. In this perspective, it is essential to identify and carefully ponder the actions to be carried out to assist and guide the ongoing revolution.

References

Uricchio, S., Parente, A.: Data driven e digital taxation: prime sperimentazioni e nuovi modelli di prelievo. Dir. prat. trib. int., 2, 921–935 (2021)

Szabó-Szentgróti, G., Végvári, B., Varga, J.: Impact of Industry 4.0 and digitization on labor market for 2030-verification of Keynes' prediction. Sustainability, **13**, 7703 (2021). https://www.mdpi.com/2071-1050/13/14/7703/htm

Chand, V., Kostic, S., Reis, A.: International - taxing artificial intelligence and robots: critical assessment of potential policy solutions and recommendation for alternative approaches. World Tax J. **12**(4) (2020)

Cordeiro Guerra, R.: L'intelligenza artificiale nel prisma del diritto tributario. Dir. prat. trib, **3**, 921–935 (2020)

Dorigo, S.: The 'algorithmic revolution': fair taxation, social pact and global governance. In: The IT Revolution and Its Impact on State, Constitutionalism and Public Law, Bloomsbury Publishing, pp. 161–179 (2021)

Farri, F.: Digitalizzazione dell'amministrazione finanziaria e diritti dei contribuenti. Riv. dir. trib. pp. 115–139 (2020)

Celotto, I robot possono avere diritti? In Intelligenza artificiale e diritto, Franco Angeli, pp. 205–216 (2020)

Oberson, X.: Taxing Robots: Helping the Economy to Adapt to the Use of Artificial Intelligence. Edward Elgar (2019). Passim

Englisch, J.: Digitalization and the future of national tax systems: taxing robots? In: Tax and the Digital Economy. Wolter Kluwer, pp. 261–281 (2019)

Lexer, M.G., Scarcella, L.: Artificial intelligence and labor markets. A critical analysis of solution models from a tax law and social security law perspective. Rivista italiana di informatica e diritto, **1**, 53–73 (2019)

Uricchio: Robot tax: modelli di prelievo e prospettive di riforma. Giur. it., 7, 1749–1761 (2019)

Contrino: Banche dati tributarie, scambio di informazioni fra autorità fiscali e protezione dei dati personali: quali diritti e tutele per i contribuenti?. Riv. dir. trib. 29 maggio 2019

Dorigo, S.: La tassa sui robot tra mito (tanto) e realtà (poca). Corr. trib. **30**, 2364–2370 (2018)

Beretta, G.: EU VAT at the dawn of the fourth industrial revolution. World J. VAT/GST Law **3**, 1–22 (2018)

Oberson, X.: Taxing robots? From the emergence of an electronic ability to pay to a tax on robots or the use of robots. World Tax J. 247–26 (2017)

Giovannini: Legalità ed equità: per un nuovo sistema impositivo. Dir. prat. trib., **6**, 2335–2369 (2017)

Fransoni, G.: Per la chiarezza delle idee su Bill Gates e la tassazione dei robot. Riv. dir. trib. 10 marzo 2017

Tanzi, V.: Fiscal policy, growth and employment. Riv Diritto finanziario e scienze delle finanze **4**, 489–512 (2015)

Autor, D.H.: Why are there still so many jobs? The history and future of workplace automation. J. Econ. Perspect. **29**(3), 3–30 (2015)

Mazur, O.: Taxing the robots 46 pepperdine law review, pp. 279–295 (2019). https://scholar.smu.edu/cgi/viewcontent.cgi?article=1369&context=law_faculty

Barros, M.: Robots and tax reform: context, issues and future perspectives, 2 Intl. Tax Stud. **6**, 2–6 (2019). Journal Articles & Papers IBFD

Frey, C.B., Osborne, M.A.: The Future of Employment: How Susceptible are Jobs to Computerisation? Oxford, Oxford Martin School (2013). https://www.oxfordmartin.ox.ac.uk/downloads/academic/The_Future_of_Employment.pdf

Oxford Economics: How robots change the world - what automation really means for jobs and productivity. June 2019. https://resources.oxfordeconomics.com/how-robots-change-the-world?source=recent-releases

Dimitropoulou, International - Scaling Back Tax Preferences on Artificial Intelligence-Driven Automation: Back to Neutral?

Social Security and Robotization

María Amparo Grau Ruiz[1,2](✉)

[1] Financial and Tax Law, Complutense University of Madrid, Madrid, Spain
graurUiz@ucm.es
[2] Transnational Taxation, Northwestern University, Chicago, USA

Abstract. This paper outlines the main problems, provides a discussion, searches for solutions and tries to offer an international perspective on the area of Social Security and robotization. This topic has been identified as a major concern for all the stakeholders. Following a sketch of the issues, some positive impacts of automation for this area are outlined.

This contribution raises awareness on the need to simultaneously address innovation for competitiveness while keeping the social impact in mind, It address both the options available to the public authorities' and the possible reactions of the companies/workers. Its proposals are aligned with the work of the International Social Security Association and comparative experiences with proven good results. The basic recommendation is to increase joint investment in reskilling to ensure employability. This could keep as many workers as active as possible, allowing flexible and multiple accesses to the Social Security system. The automatic tracking of changes in the circumstances affecting both workers and beneficiaries could serve to effectively adapt it.

1 Introduction

CONSIDERING the multisectoral expansion and the growing implementation of robots, careful attention should be paid to the possible impacts of robotics (both negative and positive) on the Social Security system, in order to try to minimize and enhance them respectively.

Individually and collectively, protective sufficiency must be sought, reinforcing intergenerational solidarity. In the face of robotization, the main social need is to invest in adequate training to reduce the vulnerability of workers at risk of losing their jobs at present, guaranteeing their future employability in case of digital transformation.

IINBOTS CSA, "Inclusive Robotics for a Better Society", EU H2020 project, G.A. No .780073.
AudIT-S, "Legal and financial significance of sustainability audit schemes through smart data management", Spanish National Research Plan, Ministry of Science and Innovation, PID2019-105959RB-I00.
XXITAX, "The necessary update of tax systems to face the challenges of the twenty-first century", Generalitat Valenciana PROMETEO project 2021–041.

M. A. Grau Ruiz (Ed.): INBOTS 2021, BIOSYSROB 30, pp. 143–148, 2022.
https://doi.org/10.1007/978-3-031-04305-5_23

2 Sustainability Concerns Affecting Both the Social Security System and the Companies

The problem that lies behind the actual discussion is that of Social Security's sustainability vs. companies' competitiveness. Robot-created risk involves both sides of the Social Security system: revenue and expenditures. The revenue perspective is affected by the number of workers and/or taxpayers contributing. The level of expenditure depends on the number of retirees and/or beneficiaries demanding coverage of their needs. When an imbalance is created due to a lower amount of revenue available to finance a growing expenditure, the stability of the social security system is seriously compromised.

A countervailing measure is reskilling [1]. Additionally, investment in human capital equals fighting inequality. However, questions remain over who in the public and/or private sector could potentially bear the financing costs of training? This is not easily solved - making robotics expensive to achieve funding would be against EU businesses' global competitiveness.

There are already cases that allow the entry or return of workers to the labor market. The existing blend of public and private support mechanisms applied to them can be useful to devise a solution for displaced workers due to robotization processes. Some of these rules for relatively similar purposes include (a) reductions of the company/worker's contribution to Social Security/funds, in cases such as senior workers, change of workplace (risk for pregnancy/occupational disease), selective employment (after partially/totally recovering capabilities), discontinuous permanent workers; or reduced contribution for insurance against accidents, if there is a reduction of risks and dangers at work; and (b) contracts for training and learning (combination of periods for education and work), or professional training for employment (where companies may program activities for their workers or give permits) [2].

An impact statement on social, environmental and/or governance matters should define the advantages and disadvantages that different possible forms of robotization bring in terms of the different tasks that the business activity comprises, detailing the possibilities of "circularizing" the economic processes and the workforce [3–5]. To explain them to the workers and other stakeholders and to debate alternatives would be beneficial in private and public organizations when introducing the use of different types of robots. This sort of dialogue should be highlighted in the current circumstances to promote acceptability of the changes proposed [6].

It is convenient for the sustainability of the system that workers have the necessary training to stay active, that their salaries correspond to the most advanced level of qualification and that their contributions grow as much as possible. In order to survive, the business sector has an interest in ensuring that the contribution is fair and that the benefits received in turn maintain their dependency relationship. Above all, in the long-term this sector may end up indirectly assuming the cost of those laid off, whether it is financed by Social Security itself or by the State through a tax increase.

Using higher taxation to "penalize" the innovative companies that use robots to be more competitive and productive is not the ideal solution. But neither can these companies ignore the need to calibrate their effects on the system that serves the environment in which they operate. They should also consider that the greater the number of people who need assistance protection, the lower their ability to purchase goods and services, which will affect the market and companies' long-term survival.

In the case of robots, on the one hand, there is a conceptual difficulty in identifying the "subject" to be forced to contribute due to their lack of personhood, regardless of who may be responsible for the payment of a new contribution. On the other hand, the use of robots can have an impact on the economic growth of a given society or State by increasing general productivity and, in particular, the competitiveness of companies, which should not be undermined. The introduction of a robots' contribution would require examining what parameters should be taken into account to determine the specific amount due to quotation (or taxation) to be established on each "robot", which, almost certainly, would be a major source of difficulties and conflicts [7].

However attractive it may seem at first glance, the line of argument that promotes charging/taxing the introduction of robots "in terms similar to the rate applicable to the jobs it replaces" or through the "introduction of specific taxes for these circumstances that allow the recovery of the losses derived from the disappearance of employment, which not only affect the Social Security systems -due to the lack of contributions, but also to all the actions of the State -due to the disappearance of wages subject to income taxation" [8], raises serious technical problems and will be practically difficult given the current scope of robotization of certain tasks (not complete jobs), which can also be modified over time.

3 An Integrative Solution Combining Principles in the Tax and the Social Security Systems

A possible solution to the aforementioned problem is to try to make clear the relationship of social contributions or tax charges directly with the way companies use the technology. In this regard, it is worth remembering that the possible fiscal or social legal treatment to be configured would not only have to be limited to the imposition of taxes but would also have to contemplate the possibility of encouraging certain desirable behaviors. In particular, the tax incentives for research, development and innovation (R&D + i) could be linked to the requirements of responsible research and innovation within the European framework, as we preach in the field of robotics to avoid incurring into a false dilemma [9].

From the perspective of the economic capacity (ability to pay principle), there is no magic algorithm for the distribution of the profits derived from robotization. Who benefits from it? Society? The company that implants the robots? The one that manufactures them? Any worker? Logically each of these parts should contribute, in accordance with some progressive scale so as to sustain the system. As this is not an easy solution, the focus is changed to the extra-fiscal mechanisms (also named regulatory taxation), which take into account the negative and/or positive externalities generated to adjust the tax burden. This could happen in the field of employment and in the social protection system.

Nonetheless modifying a Social Security model to decouple it from employment would mean changing those responsible for its financing, who would cease to be employers and workers through contributions, to impose the burden on the citizens via taxes. In cost/benefit terms, this could be interpreted as a tax relief for employers, who would stop assuming the most significant part of the financing of social protection, and a burden for society, but fundamentally for workers. They would start to assume a substantial part of the financing of their own social protection via taxes (especially if revenue obtained from taxes on consumption is dedicated to social protection).

But first, the possibility of continuing to maintain contributions according to economic capacity in the ambit of the Social Security should be seriously reviewed. What does not seem clear is that, in order to quickly discharge the financial pressure on the Social Security system, by altering its structure, the contributory features (that basically define the equivalence between contributions and benefits) are transferred to the tax system.

When shifting financing certain costs and requiring the State based on the efforts of all citizens to face them is proposed, a limited vision of the tax system reduced strictly to taxes is often adopted. This overlooks the fact that together with the economic capacity, the principle of benefit is also taken already into account in this area (for example, in fees and special contributions when a good or service is received as a consideration).

The solution cannot be found by changing the problem of location, but by aiming at optimally combining these two principles (ability to pay and benefit) that must play a decisive role in both systems together (the tax system and the Social Security system). In the end the actors in both fields are identical and there are no more resources in one than in the other (it is a zero-sum game). The key is to set the criteria for burden sharing well, promoting cooperation to achieve maximum joint benefit.

4 An Approach Coherent with the Proposals Made by the International Social Security Association

The International Social Security Association (ISSA) has made studies in line with what has been previously suggested here. It has observed a recent trend of including an increasing percentage of workers (even if they have incomplete work histories) in contributory regimes, making the eligibility threshold more flexible. As mentioned, the more contributors there are, with contributions that are more adjusted to their reality, the more equitable and sustainable the system will be.

In the context of robotization, pooling risks is convenient for employers, because it would function as a kind of insurance against short-term and cyclical crises. The ISSA warns that the "relationship between the dismissal contribution and the social costs thereof is the one that can best reflect the total cost of replacing workers with artificial intelligence and robots" [10]. Reference is made to the US *bonus malus* mechanism for unemployment insurance, which sees higher percentage of the social cost associated with the lack of employment granted to those who generate the cost, with positive consequences in terms of reducing the total burden of the contributions and increases in employment [11].

At the same time, ISSA assesses other measures, that could be useful in the current context, such as the use of broader income ranges to determine contribution levels, the deferral of contributions during economic downturns, or the interruption of contribution periods. It highlights the possibility of providing temporary financial support to the self-employed (a supplement to reach the usual level of social assistance or an interest-free loan -as in the Netherlands). In short, ensuring workers remain active as long as possible is imperative (even if this implies specific sacrifices in the regular collection of contributions). It is evident that an interest-free loan has a lower economic cost than the increase in the list of unemployed.

5 Remark on Positive Impacts of Robotics

On the one hand, it is evident that the continuous capture of data by robots can serve the purpose of correctly accommodating the Social Security system in real time [12], both from an individual and collective perspective. A greater adaptation of the quotas to reality, due to information being available practically instantaneously, is now technically feasible (for example, with wearable robots or exoskeletons and interactive robots facilitating collaboration in tasks carried out and/or supervised for people). This is an obviously positive outcome of robotics.

On the other hand, the working life of the worker could be prolonged by avoiding certain injuries to their health thanks to the use of the robot (e.g., caused by loads, repetitive tasks, etc.). This would mean cost savings in benefits derived from occupational diseases and the possibility of reaching a longer contribution period.

6 Conclusion

Employability reduces pressure on Social Security systems. Reductions in Social Security contributions may help in robotics transition (a) if clear requirements for reskilling are set; and (b) proportionate benefits are controlled.

An EU-wide solution would be desirable approaching legislation or at least co-financing. This should also account for positive impacts.

References

1. Ruiz, M.A.G.: INBOTS Project report. Concerns and expectations raised by robots: some experts' opinions to guide policymakers. Dilemata **35**, 100–103 (2021)
2. Díaz, A., Ruiz, M.A.G.: Social Security and robotization: Possible ways to finance human reskilling and promote employment. Paladyn J. Behav. Rob. **11**(1), 340–350 (2020)
3. Ruiz, M.A.G., O'Brolchain, F.: Environmental robotics for a sustainable future in circular economies. Nat. Mach. Intell. **4**, 3–4 (2022)
4. Ruiz, M.A.G.: Razones para la tributación de la robótica debidamente enmarcada en la economía circular" [Reasons for a taxation of robotics duly framed in the circular economy]. In: Vivel-Búa, M., García Novoa, C. (Dirs.) Digitalización, inteligencia artificial y economía circular [Digitalization, AI and circular economy], pp. 51–63. Thomson Reuters Aranzadi, Cizur menor (2021)
5. Ruiz, M.A.G.: Medidas fiscales para que la robótica contribuya el desarrollo sostenible? [Fiscal measures to allow robotics contribution in favor of sustainable development?]. In: Serrano Antón, F. (Dir.) Inteligencia artificial y administración tributaria. Eficiencia administrativa y defensa de los derechos de los contribuyentes [AI and tax administration. Administrative efficiency and defence of taxpayers' rights]. Thomson Reuters Aranzadi, Cizur menor (2021)

6. Goos, M.: A skilled workforce – ready to contribute to tomorrow's world of work. In: Ruiz, M.A.G., et al. (eds.) INBOTS White Paper On Interactive Robotics Legal, Ethical & Socioeconomic Aspects, pp. 130–134 (2021). http://inbots.eu/wp-content/uploads/2021/09/D.2.2-FINAL-INBOTS-White-Paper-WP2_-Legal-ethical-socioeconomic-aspects-.pdf
7. Álvarez, A.L.: Una Seguridad Social para el futuro, con un apunte sobre automatización, robots y Seguridad Social [A Social Security for the future, with a note on automation, robots and Social Security]. In: Sánchez-Urán Azaña, M.Y., Ruiz, m.a.g. (Dirs.) Nuevas Tecnologías y Derecho [New Technologies and Law], p. 135. Juruà, Porto (2019)
8. Peset, J.M.G.: Digitalización, robotización y protección social. Revista Teoría y Derecho **23**, 125 (2018)
9. Ruiz, M.A.G.: El actual dilema fiscal por el impacto de la robótica: innovación o empleo?" [The current fiscal dilemma due to the impact of robotics: innovation or employment?]. In: Torralvo, J.J.H., Cruz Padial, I. (Dirs.) Tributos, servicios digitales y robótica [Taxes, digital services and robotics], pp. 327–346. Thomson Reuters Aranzadi, Cizur menor (2020)
10. La Salle, D., Cartoceti, G.: La Seguridad Social en la era digital, cómo pueden los sistemas de seguridad social aprovechar las nuevas oportunidades y afrontar los nuevos desafíos, AISS, Geneva, p. 21 (2019)
11. Sánchez-Urán Azaña, M.Y., Ruiz, M.A.G.: Economía Digitalizada y Seguridad Social. Perspectivas jurídicas y económicas del "Informe de Evaluación y Reforma del Pacto de Toledo". In: Hierro, F.J.H. (Dir.), pp. 905–946. Thomson Reuters Aranzadi, Cizur menor (2020)
12. Ruiz, M.A.G.: Robotics, artificial intelligence & big data shaping fair taxation for sustainability in the context of digital economy. In: Rayo, A.O., Aucejo, E.A. (eds.), Global Tax Governance: Taxation on Digital Economy Transfer Pricing and Litigation in Tax Matters (MAPs + ADR) Policies for Global Sustainability. Ongoing U.N. 2030 (SDG) and Addis Ababa Agendas, pp. 187–198. Thomson Reuters Aranzadi, Cizur menor (2021)

Pensions, Robotics and Taxes, the New Challenges of Taxation

Miguel de Haro Izquierdo(✉)

Financial and Tax Law at Faculty of Law of the Universidad Complutense of Madrid, Madrid, Spain
migueldh@der.ucm.es

Abstract. Robots will change countries' traditional Social Security and tax systems. Then, the question is, how can such changes be integrated in our societies and legislation?

1 Introduction

In the following work we address questions that directly affect the future of the welfare state with the emergence of AI and robotics.

Should robots be considered persons? Will they put an end to the Will robotisation generate greater among citizens? Should they pay Social Security contributions as robotised workers?

Robots will change countries' traditional Social Security and tax systems. Then, the question is, how can such changes be integrated in our societies and legislation?

2 Wellfare System Crisis

In the most developed economies, major technological and demographic changes are taking place and are having a decisive influence on governments' ability to obtain the economic resources with which to meet the costs of welfare systems. Demographics is an essential element in pension financing systems The challenge of social welfare systems is focused on collecting taxes, balancing public finances, and maintaining a very complex balance of pensions and health services. The main direct consequences of population statistics will have a decisive influence on the income of the elderly population in the coming years and will be influenced by:

1.- Decrease in the population due to the low birth rate.
2.- Increased life expectancy of both men and women.
3.- Significant reduction of the economically active population.

At the same time, we must emphasize that the consequences of the distribution of the future population will generate a series of socioeconomic effects that can be synthesized in the following aspects:

M. A. Grau Ruiz (Ed.): INBOTS 2021, BIOSYSROB 30, pp. 149–154, 2022.
https://doi.org/10.1007/978-3-031-04305-5_24

1.- Reduction in the number of students and active people.
2.- Reduction of contributions to sustain the high rates of needs of the elderly.
3.- Increase in the economic cost on pensions, health, infrastructures and dependency.
4.- Reduction of the tax collection capacity, which will make it difficult to balance public finances or maintaining a very complex balance between pensions and health services.

The above population description and its most immediate consequences mean that the pension systems of the OECD countries must face the challenges created by the high rate of population aging, demographic, social, labor, macroeconomic and financial risks. In this situation, it is necessary to address proposals that will allow an immediate solution to be found in the essential bases of the welfare system.

3 The Contribution of Robots to Pensions

The fourth industrial revolution will minimise career opportunities for a large part of the population and will also have an impact on economic aspects of public spending policies of states, consumption, reduction of welfare state income for pensions and services for the elderly, the payment of subsistence income, etc.

Nowadays, the world economy has a serious deficit of workers with tertiary education, and a deficit of workers with some kind of qualification to respond to the needs of the labor market, which total more than ninety million working people. At the same time, the need for qualified personnel equals the number of workers who do not have any professional qualifications. This situation generates distortions and inequalities among the population in terms of income and social integration. From the point of view of state income, these inequalities can generate a greater volume of income for those who are more qualified and an expulsion from the labor circuit to work in the informal economy, which generates inequality and, above all, a deficit in state revenue collection. Economists predict an influence of robots on labour recruitment, on the availability of labour and a decrease in wages, savings, pension contributions, so that there will be a shift of resources from labour to capital.

We do not believe that the reduction in Social Security contributions can be replaced by a contribution from robots to Social Security as such, as this would distort the very essence of the concept of contributions, as these are benefits and measures that are put in place by Social Security to anticipate, repair or overcome certain situations of misfortune or states of need of workers, and therefore tend to cause a loss of income or excess expenditure for the people who suffer them. We understand, *a priori*, that it is not possible for the robot to obtain these benefits as if it were a legal person with rights and obligations.

Being subject to Social Security obligations and receiving benefits is intrinsically linked to the fact of recognition of a legal personality, consequently, the capacity of rights and duties is the only attribute that is exclusively considered by the courts to determine the legal personality of any entity.

If robots, with an industrial and social character, have been enabled to do different kinds of acts by humans, it is impossible that they can be granted a legal personality.

Robots, as well as all elements of AI, are treated as tools is legally responsible for their actions, responsible for ensuring that they operate within the limits of the law.

Autonomy and self-determination are not considered grounds for legally conferring rights over any entity, and thus do not manifest a legal standing as a subject of rights and obligations. Being subject to Social Security obligations and receiving benefits is intrinsically linked to the fact of recognition of a legal personality, consequently, the capacity of rights and duties is the only attribute that is exclusively considered by the courts to determine the legal personality of any entity.

4 Some Tax Measures That Can Complement the Reduction in Income from the Spread of Automation

Tax policy has a key role to play in tackling inequality, including income inequality that may be increased as a result of the social revolution and robot work.

In this way, we believe that tax policy plays a key role in the fight against inequality, Shifting taxation from labor income to a higher tax burden on capital income could mean a way to redistribute revenue. The favorable points of this proposal are that it would allow to maintain a greater volume of income to citizens and families, although we must recognize that it could also generate an important gap in economy since the possible development of new investments and attraction of capital would be paralyzed by the fiscal measures that would become a punishment.

Establishing Social Security contributions based on the productivity of robots as opposed to replacing labour income would alleviate the reduction in revenue from labour contributions.

As a corrective measure, parallel to the previous ones, a universal basic income could be established by granting a fixed and universal amount to all citizens, regardless of their income level and behavior. This model has been experienced locally through social programs, such as the one developed in Nordic and North American countries. As we indicated, a basic income could be established for those persons who have been expelled from the labor market as a consequence of the technological revolution or have been excluded from the labor circuit as a consequence of a lack of technological adaptation or amortization of their jobs as a consequence of the artificial intelligence revolution.

The major drawbacks deriving from the establishment of a basic income are, the difficulty of maintaining a budgetary balance in ensuring this income; the social incentive and the effects it may generate, from the human point of view, and the integration of people in society without a link to work and supported by the State. The lack of relationship of large sectors of society with the world of work and social integration would generate great inequalities, not only economic ones. The corrective element of this minimum income would be complicated to implement if it were implemented in a generalized manner among the population, due to the enormous volume of resources to be financed in economies with a particularly rigid budgetary discipline and conditioned by supra-state decisions.

The third of the measures, to be taken into account, is the inequality generated by the possible decrease in access to sources of income in front of the fact of the labor market robotization, as well, it should be considered that fiscal policy plays a fundamental role in the fight against inequality. To this end, it would be necessary to define policies to maintain Social Security income as a guarantee of the commitments assumed by the welfare state.

In this section, it should be determined whether it is feasible to transfer sources of revenue to capital income if there is a drastic drop in labor income due to the replacement of people by robots. In a report by the World Bank, the Industrial Revolution 4.0 is defined by very clear characteristics, in which the robot and artificial intelligence are the central axis of change.

The technological revolution will create an clare displacement of the influence of robots in labor hiring, in the disposition of the labor force and, especially important will be the decrease that it will cause in wages, savings, contribution to pension plans, both public and private, so that there will be a transfer of resources from labor to capital.

The revolution of robotics and artificial intelligence brings up a need to reorder the personal income of dependent workers, not only from labor but also from capital, as well as an assessment and study of the income from social contributions and payment to the maintenance of the welfare state.

The great challenge that will have to be faced socially is the one that derives not only from the reordering of work by robots, but also from the drastic demographic fall of the European continent and especially Spain. The differences of a depleted active population contributing resources to Social Security, and on the other hand, the high economic costs of the elderly and dependent "consumers of state resources" to which pensions must be provided, will generate higher health care costs and significant allocations of resources for dependency that may lead to serious financing problems, especially in social security contributions.

In Spain, the social security system is supported by contributions that are assumed in a proportional part by the employer and in another part by the worker, with the State subsidizing, on certain occasions, the system of social contributions through the Budget Law. The fundamental source of financing of the pension system is determined by the contribution quotas, which are assumed by the employers to cover the risks derived from contingencies of the workers.

It is not plausible to create a contribution from robots by means of Social Security as such, since this instrument would be disfigured in its own essence as a contribution concept, since these are benefits and measures that put Social Security into operation to foresee, repair or overcome certain situations of misfortune or states of need that usually originate a loss of income or an excess of expenses in the people who suffer them. We understand that it is not possible for the robot to obtain these benefits as if it were a legal person with rights and obligations.

The clearly positive economic results of the new technification, undoubtedly, have an effect on the increase in productivity. It is only fair that there would be established a tax compensation for the generation of resources and the increase in industrial productivity, or in the services of the companies provided by the robots, not by means of social security contributions, but through alternative means such as the establishment of a basic income, distribution of the obtaining of state income by transferring a lower collection of personal labor income to capital income.

The World Economic Forum on the European Union indicates that;

First, labor income generates inequality by gradually increasing as low-income people are left behind in their incorporation into the labour market; this is the main cause of income inequality in the population.

Second, capital income and wealth are increasingly concentrated at the top of the population distribution, exacerbating total income inequality as the share of capital in GDP rises.

Finally, although fiscal policy has a large offsetting effect on inequality in EU countries, it is increasingly difficult to offset labor inequality relative to capital income. It seems that inequality in the labor market, and we understand that it is perfectly extrapolated to a situation of labor deficit in the era of artificial intelligence, generates greater inequality and economic imbalance in the population.

As a corrective measure for the above, contributions to Social Security could be established based on the productivity of robots as opposed to the substitution of labor income. Capital income could be distributed among those who have lost their jobs or have not adapted to the change. The transfer of income to capital could be a clear brake on investment and the economy, but if it is established gradually and based on robot productivity parameters, new revenue-raising innovations could be introduced that would not harm investment.

Finally, and according to models of certain countries, such as the United States through the Alaska model or Sweden, the revenue generated by the robots could be allocated to large investments in sovereign wealth funds that would manage this income by generating economic resources and their results could be distributed among those with lower incomes or contributors to the Social Security system.

5 Conclusion

The great challenge we face is that arising from the reorganization of work by robots. The differences between an increasingly depleted active population contributing resources to Social Security and, on the other hand, the high economic costs of the elderly and dependent "consumers of state resources" for pensions, will generate an increase in health expenditure and significant allocations of resources for dependency that could lead to serious financing problems. Taxation, as a redistributive tool, can generate a tax-based compensation for the generation of resources and the increase in industrial productivity, or in the services of highly robotised companies, not by means of Social Security contributions, but through alternative channels such as the distribution of state income by transferring less revenue from personal labour income to capital income, or, as a more recommendable measure, the establishment of contributions to Social Security according to the productivity of robots with a clear purpose of organising income.

References

1. Deloitte Services LP. Forces of change: Industry 4.0 A. Deloitte series on Industry 4.0
2. Ray, D., Mookherjee, D.: Growth, Automation and the Long Run Share of Labor (2019)

3. Solaiman, S.M.: Legal personality of robots, corporations, idols and chimpanzees: a quest for legitimacy. University of Wollongong (2010)
4. Sachs, J.D., Benzell, S.G., LaGarda, G.: Robots: curse or blessing? a basic framework. Working Paper 2109. National Bureau of Economic Research (2015). http://www.nber.org/papers/w2109
5. Obama, B.: Intelligence, Automation, and the Economy. Executive Office of the President (EOP) (2016)
6. Eurostat Statistics Explained. "Population and population change statistics" (2019)
7. OECD Pensions Outlook. OECD Publishing (2018)
8. Understanding Changes in Inequality in the EU Background to "Growing United: Upgrading Europe's Convergence Machine". World Bank Report on the European Union

Taxation and Robotics in the Recent EU Context: Should My Dishwasher Contribute to Repay the Post-COVID-19 Debt?

Franco Roccatagliata[1,2](✉)

[1] EU Tax Law, College of Europe, Bruges, Belgium
franco.roccatagliata@coleurope.eu
[2] European Commission, Brussels, Belgium

Abstract. Nowadays, robotics and artificial intelligence are an important aspect of our lives. They are a source of wealth but, in the short run, they do not generate new fiscal resources. If tax systems do not adapt to these new technologies, fiscal resources could even be lost. Can we intervene in a way that taxation facilitates scientific progress without hindering its development? Bill Gates proposes a controversial solution that could be added to the list of the potential resources useful for repaying the public debt caused by the COVID-19 pandemic. Here the issue is not analyzed from a technical point of view, but rather with a more limited perspective. In specific, to relay the well-known idea of a robot tax alongside another current topic which is the economic effect of the pandemic.

1 Introduction

"I'*VE seen things you people wouldn't believe.* Robots moving their tax residence *off the shoulder of Orion… I watched C-beams* turn *in a flash* in hybrid entities just to go through *Tannhäuser Gate* untaxed. *All those moments will be lost in time, like tears in rain…* Time to pay taxes. *Time to die*".

The views expressed in this article are those of the author and cannot be attributed to the European Commission.

M. A. Grau Ruiz (Ed.): INBOTS 2021, BIOSYSROB 30, pp. 155–160, 2022.
https://doi.org/10.1007/978-3-031-04305-5_18

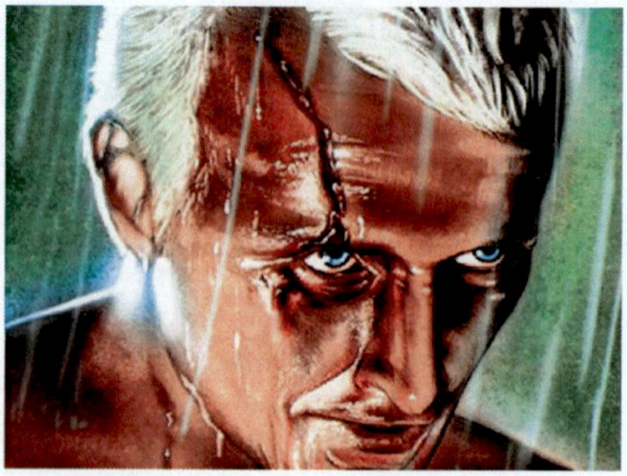

Roy Batty (portrayed by Rutger Hauer), the rogue artificial 'replicant' in *Blade Runner* of Ridley Scott (1982)

The *Blade Runner*, a cult movie, ends with the monologue "*Tears in rain*" performed by a human-like robot, [1] one of the most famous monologues in the history of cinema. Maybe, the next episode of the *Blade Runner saga* could end with the variation proposed above: the replicant struggling - like us, humans - with insoluble fiscal problems and only two certainties in the real world: death and taxes [2].

But, why should robots (or any other automatic home appliance including … a dishwater [3]) pay taxes and/or try to evade them? Let's go in order, starting by defining the object of our intervention.

2 COVID-19 Pandemic and Economic Crisis

As we know, COVID-19 pandemic is not only a sanitary emergency. It is, for example, a drastic change of our work habits. It is, overall, a real economic disaster. Unfortunately, taxation - in the meaning of ***public revenue*** - is no exception. We expect that the current crisis will dramatically reduce the tax revenues in the near future. The OECD, the European Parliament and other institutions have pushed to explore altogether a new option for tax policy.

In a recent study [4] the OECD - analyzing this pandemic from an economic point of view - has identified four different phases.

1) In an initial phase of the crisis, public budgets supported measures in favor of healthcare structures and the fight against pandemic with vaccines and medical equipment.
2) Then, budget intervened in favor of business cash-flow, employment, and even family's income affected by the pandemic.
3) As third step, public resources are intervening (or will intervene soon) to stimulate economic recovery.

After these three phases, we will land pretty soon in a new fourth phase:

4) Restoring public finance.

Therefore, doing that will require a lot of creative imagination!

Many solutions are already proposed to pay off the debt generated by Covid-19 crisis. Among the various solutions, one is to mutualize the cost of the pandemic in Europe: the issuance of Euro/Coronabonds, the creation of an EU coronavirus rescue funds, etc. [5].

In particular, the cost of that crisis has raised awareness of temporary recovery tax contributions as a possible source of finance in the pandemic aftermath.

Some economists and politicians called for a ***temporary wealth tax***. It is probably the easiest solution to apply. Prof. Gabriel Zucman, the director of the recently created *European Tax Observatory* (*EU Tax*) - thanks to EU funding – presented this policy option in various papers [6, 7].

The Secretary General of the United Nations, Mr. Guterres, proposed to introduce a ***solidarity tax*** [8]. A tax to reduce extreme inequalities that should be paid by those who have profited the most during the pandemic.

In a letter to various Governments, even 84 multi-millionaires (***Millionaires for humanity***) have asked to use their fortune in order to finance the post-COVID economic recovery [9]. Their message concludes with an incredible cry: *So, please, tax us, tax us. It is the right choice; it is the only choice.* At least this is not usual.

In the EU, some members of the European Parliament proposed to finance the fight against COVID-19 and the economic recovery with an effective ***financial transaction tax*** (FTT). Unfortunately, FTT is a sensitive issue; in the Council of the EU, it has been repeatedly rejected in the recent past: a debate on an EU Financial Transaction Tax started more than twelve years ago and, until now, it didn't bear any fruit.

Other members of the Parliament are pushing for a common ***EU tax on digital services*** [10]. In 2018, the European Commission presented a proposal on the same subject [11] but, at least for the moment, discussions in the Council have been 'freezed' waiting for positive developments from Paris. In fact, in OECD premises, the 141 members states (and jurisdictions) of the *Inclusive Framework* - practically the whole World - are still discussing on a global proposal (so called *BEPS 2.0* or *Two-Pillars* Proposal) for the ***taxation of the digital economy*** and a ***minimum effective tax*** on profits of multinational enterprises [12]. The European initiative might return to the table of the Council in case of unexpected failure of this negotiation in Paris. However, even if the original timing (2020) will not be respected, it seems probable that the general political agreement recently reached will be operative in 2023.

In a deal on the EU's 2021–2027 budget on 10 November 2020, European Institutions reached an agreement on a binding timeline for the introduction of new EU revenue sources (not totally identified yet). These new revenue sources would help paying off the joint debt taken by EU countries to finance the COVID-19 recovery. Without new own resources, the borrowed recovery money would have to be paid back through further reductions to EU programs and/or higher GNI-based contributions from EU countries.

At the moment the proceeding has made its first steps but in this moment it is important to ensure that the burden will be not on *the usual suspects* - employees, SMEs, etc. - but on the tech giants, tax dodgers, big foreign polluters and others who do not currently pay their fair share [10].

Let me add the *robot tax* at the end of that long list.

3 Robot Tax as Contribute to Repay the Post-COVID-19 Debt

The other speakers of this session on *Taxation & Robots* in the *INBOTS* Conference have explained - in this occasion and on many leading scientific contributions [13, 14, 15, 16] - how a robot tax can work. How, eventually, implement it. But also, that a new tax on *Artificial Intelligence* (*AI*) would not solve all budgetary difficulties and, if not well structured and attentively reflected, it can even generate new problems. This final intervention of the session has not any pretention to explain the issue from a technical point of view.

The ***robot tax*** is an idea launched, some years ago - among others - by someone very familiar with high technology: Bill Gates.

Mr Gates believes governments should tax companies' use of automatic processes that are taking human jobs, as a way to slow the spread of automation and take time for requalification of workers. According Mr Gates' view, "*Right now, the revenue of human workers is taxed ... If a robot comes in to do the same thing, you'd think that we'd tax the robot at a similar level*" [17].

Bill Gates, founder of Microsoft, business magnate and philanthropist.

The *robot tax*'s idea developed in this article is not different then from Mr Gates' proposal or from similar older proposals: any extra-productivity related with automation should generate more taxes or, at least, not less than before.

Comparing this suggestion with previous ones, the only element of novelty is on the purpose of the measure.

It would be a contribution - a *robot tax* (or, maybe better: an ***automation tax***) - not directly related to the impact of the automation and robotics on social policies and social security welfare, but simply a new potential resource for the national *Treasury* to respond to economic challenges generated by COVID-19 pandemic.

Who would pay this tax? The robots? Certainly not.

We are far from attributing a *legal personality* to a machine and *Star Wars' C-3PO* and *R2-D2 droids* have not fulfilled a tax return yet.

Logically, it would be the ***beneficial owner*** of the activities of the machine, particularly in cases where automation has generated profits during the pandemic.

In case of cross-border activities, eventually the robot could be considered as a ***permanent establishment*** of the *beneficial owner* for the allocation of taxing rights, according to the classic criteria adopted by the international tax law.

It would be an anachronistic tax, or could it, maybe, a barrier against the scientific progress? It difficult to say.

Robot tax has received some strong criticism. Someone said that the robot tax remembers the history of the stocking frame. William Lee developed the stocking frame in 1589 to alleviate the time-consuming labor required to produce knit caps. Lee sought a patent from Queen Elizabeth I, who rejected it, observing that the invention threatened knitters and "*would assuredly bring them ruin by depriving them of employment, thus making them beggars.*" [18].

Let us consider an example.

The most automatized country in the world (Republic of Korea) have introduced a sort of robot tax (more correctly, they simply progressively reduced the automation tax credit) without any significant impact on the technological development [19]. So, probably, the provisional answer to the question; "is a robot tax a barrier against progress?", is no.

In any case, it should be noted that at European level, many other preliminary questions are still open. And, without unanimous answers to these questions, to build a European robot tax seems, to say the least, problematic.

First of all, a definition of AI or robot. The European Parliament, in its Recommendation of 2017, suggested that a *smart robot* should be characterized *by the acquisition of autonomy ... exchanging data with its environment ...*; *at least a minor physical support ... and absence of life in the biological sense* [20].

More recently, in April 2021, the Commission published its *AI package*, proposing new rules and actions to turn Europe into the global hub for trustworthy AI. In particular, the Commission's proposal, at article 3, defines 'artificial intelligence system' as *software that is developed with one or more of the techniques and approaches ... and can, for a given set of human-defined objectives, generate outputs such as content, predictions, recommendations, or decisions influencing the environments they interact with* [21].

4 Conclusion

European governments have immediately reacted to the COVID crisis and are now discussing ways to mutualize the cost of the epidemic.

Many proposals are on the desk of the Council of the EU and on the desks of ministries of Finance of EU Member States. Some MEPs and academics (including the author of this note) consider a new tax on AI and/or on automatic systems a proposal to take into account and to add this item to the list of resources useful for restoring public finance.

However, among the various proposals that the European Institutions prepared on artificial intelligence and robotics, there are no specific document on taxation. So, it seems that it is not yet time for a European robot tax.

Death and taxes are still a problem for humans but, at least for the moment, not for *C-3PO* and *R2-D2*...

References

1. Abbott, J.: The 'monster' reconsidered: blade runner's replicant as romantic hero. In: Extrapolation, vol. 34, no. 3 (Winter), p. 340. The Kent State University Press (1993)
2. Franklin, B.: Letter to Jean-Baptiste Le Roy (1789)
3. Falcão, T.: Should my dishwasher pay a robot tax? Tax Notes Int. **90**(12), 1273 (2018)
4. OECD: Centre for Tax Policy and Administration, Tax and Fiscal Policy in Response to the Coronavirus Crisis: Strengthening Confidence and Resilience. Updated 19 May 2020
5. Wyplosz, C.: So far, so good: and now don't be afraid of moral hazard. In: Baldwin, R., Weder di Mauro, B. (eds.) Mitigating the COVID Economic Crisis: Act Fast and Do Whatever It Takes, p. 29. CEPR Press (2020)
6. Saez, E., Zucman, G.: Progressive wealth taxation. Policy Focus **19**(3), 6 (2021)
7. Klemm, A., Hebous, S., Michielse, G., Nersesyan, N.: COVID-19 recovery contributions. International Monetary Fund, p. 4, 16 April 2021
8. Guterres, A.: Speech at the economic and social council forum on financing for sustainable development, 12 April 2021. See also, United Nations, Our Common Agenda. Report of the Secretary-General, New York (2021)
9. Togoh, I.: 'Please Tax Us': dozens of millionaires urge governments to tax the rich to pay for coronavirus impact. Forbes, 13 July 2020
10. European Parliament, Committee on Economic and Monetary Affairs: Report on digital taxation: OECD negotiations, tax residency of digital companies and a possible European Digital Tax, Plenary sitting, doc. A9-0103/2021, 30 March 2021
11. European Commission: Proposal for a Council directive on the common system of a digital services tax on revenues resulting from the provision of certain digital services; COM (2018) 148 of 21 3 2018
12. OECD: Two-pillar solution to address the tax challenges arising from the digitalisation of the economy, Paris (2021)
13. Oberson, X.: Taxing Robots. Edward Elgar (2019)
14. Oberson, X.: Robot taxes: the rise of a new taxpayer. In: Bulletin for International Taxation, IBFD, vol. 8, p. 370 (2021)
15. Grau Ruiz, M.A.: Riesgos y oportunidades en la creciente digitalización fiscal. Revista Técnica Tributaria **130**(3), 7 (2020)
16. Fosch Villaronga, E., Grau Ruiz, M.A.: Expert considerations for the regulation of assistive robotics. A European robotics forum echo. Dilemata **30**, 149–169 (2019)
17. Delaney, K.J.: The robot that takes your job should pay taxes, says Bill Gates. Quartz, 17 February 2017
18. Weissman, W.H.: Why robot taxes won't work. In: State Tax Notes, p. 125, 9 April 2018. For a detailed analysis and literature review, see V. Chand, S. Kostić and A. Reis. Taxing Artificial Intelligence and Robots: Critical Assessment of Potential Policy Solutions and Recommendation for Alternative Approaches. World Tax Journal, 4, 2020, p. 711–761
19. Kovacev, A.: Taxing dilemma: robot taxes and the challenges of effective taxation of AI, automation and robotics in the fourth industrial revolution. Ohio State Technol. Law J. **16**(1), 182 (2020). see also, G. Bottone. A Tax on robots? Some food for thought. MEF - Ministero dell'Economia e delle Finanze, DF Working Papers, September 2018
20. European Parliament: Report with recommendations to the Commission on Civil Law Rules on Robotics; Rapporteur: Mady Delvaux, Plenary sitting, pp. 8–9, doc. A8-0005/2017 of 27 January 2017
21. European Commission: Proposal for a Regulation of the European Parliament and the Council layng down harmonized rules on artificial intelligence and amending certain Union legislative acts; doc. COM(2021)206 of 21 April 2021

Corporate Social Responsibility and Sustainability

Technological Change and Regulatory Protection of the Autonomy of People with Disabilities

Mario Toboso[1](✉), Manuel Aparicio[2], Ricardo Morte[3], Aníbal Monasterio Astobiza[3], Txetxu Ausín[4], and Daniel López[4]

[1] Institute of Philosophy, Spanish National Research Council (CSIC), Madrid, Spain
mario.toboso@csic.es
[2] Faculty of Philosophy, University of Murcia (UMU), Murcia, Spain
manuel.aparicio@um.es
[3] Faculty of Philosophy, University of Granada (URG), Granada, Spain
ricardo63@autistici.org, amastobiza@ugr.es
[4] Institute of Philosophy, Spanish National Research Council (CSIC), Madrid, Spain
{txetxu.ausin,daniel.lopez-castro}@cchs.csic.es

Abstract. In recent years, technological change has been very significant, including the field of assistive technologies for disability, aimed at promoting the autonomy of people with disabilities. This technological change (in which we mainly consider interactive robotics devices incorporating artificial intelligence tools) may bring with it an equally notable change in the exercise of personal autonomy, putting at risk the regulatory protection that underpins it. We refer to the repercussions that emerging technologies may have on privacy, data integrity and data protection. In this paper we will address these issues, taking as a reference the normative framework of the UN *Convention on the Rights of Persons with Disabilities*.

1 Introduction

Personal autonomy, understood as the possibility of self- government, without illegitimate interference, to decide and execute one's own life plan, is one of the defining characteristics of the modern subject [1], being the basis of the requirements for the validity of the social contract. However, personal autonomy does not operate in a vacuum; it is possible due to ethical-legal protection by means of reciprocally recognised human rights (civil and political; economic, social and cultural; third generation). Many of these rights require socio-technical mediations in order to be effective [2]. For this reason, technology must be conceived as an enabling mediation of autonomy (if I decide to communicate with a distant relative, the mobile phone will allow me to exercise that decision); also as a conditioning factor of the human rights that protect that autonomy (the mobile phone extends the right to be able to exchange information).

M. A. Grau Ruiz (Ed.): INBOTS 2021, BIOSYSROB 30, pp. 163–167, 2022.
https://doi.org/10.1007/978-3-031-04305-5_28

People with disabilities have traditionally been absent from these considerations, as their inability to exercise autonomy and, consequently, to take part in the social contract, has been assumed [3]. Today, at least on paper, the idea of personal autonomy is also attributed to people with disabilities. This change is the result of the struggle of social movements for the rights of people with disabilities, mainly the so-called Independent Living Movement. An incomplete change of paradigm has led from the medical-rehabilitative model to the social model of disability, which understands that this condition is the result of the interaction of bodies with particular characteristics with socio-technical environments that exclude them [4]. The change has not yet been completed, at least in terms of social discourse and society's perception of disability [5].

However, the social model has made steady progress in the normative field. We refer to its validity in the 2006 UN *Convention on the Rights of Persons with Disabilities* [6], hereafter "the Convention", the international normative framework from which the human rights of persons with disabilities are promoted worldwide. The Preamble to the Convention recognizes that "...disability is an evolving concept and that disability results from the interaction between persons with impairments and attitudinal and environmental barriers that hinders their full and effective participation in society on an equal basis with others", and in Article 2 the following definition is established: ""Discrimination on the basis of disability" means any distinction, exclusion or restriction on the basis of disability which has the purpose or effect of impairing or nullifying the recognition, enjoyment or exercise, on an equal basis with others, of all human rights and fundamental freedoms in the political, economic, social, cultural, civil or any other field". This international normative framework, knowledge of which should be necessary for anyone whose work revolves around disability, includes important references to technological developments, in relation to the principle of universal accessibility.

Section (v) of the Convention's Preamble emphasizes "the importance of accessibility to the physical, social, economic and cultural environment, to health and education and to information and communication, in enabling persons with disabilities to fully enjoy all human rights and fundamental freedoms". Accessibility is one of the "General Principles" (Article 3) guiding the Convention, and its entire Article 9 is devoted to this subject.

Palacios and Romañach [7] proposed that a new perspective is needed—a new model of disability for society to consider that is based on a definitive acceptance of human diversity and rises above the ability/disability conceptual dichotomy. This new approach, which its authors call the 'diversity model', is presented as an evolution of the social model. Whereas, in the social model, the basic principles are a person's ability and opportunity to make a contribution in society, the basic principles of the diversity model are diversity and human dignity. Therefore, it goes beyond the core theory of ability which, under the former models, imposed a vain attempt to 'be like the other person'—to aspire to a normality that is statistically incompatible, in many cases, with the intrinsic diversity of human beings. With the concept of "functional diversity", a new view was proposed that is not negative and affirms the fact that it refers to people who have With the concept of functional diversity, a new view was proposed that is not negative and affirms the fact that it refers to people who have potential access to the same functions as the rest, but in a different way.

Reinterpretation of disability as a form of human diversity is making significant inroads in some international initiatives, very prominently in the General Principles of the Convention (Article 3): "Respect for difference and acceptance of persons with disabilities as part of human diversity and humanity", and in the *Report of the Special Rapporteur on the rights of persons with disabilities*, Catalina Devandas-Aguilar, submitted to the UN Human Rights Council [8].

2 Materials and Methods

The Convention, as the international normative framework that defines and protects the autonomy of persons with disabilities, as well as the bibliography listed at the end of the paper, are used as working materials.

We adopt philosophical-practical methods based on: analytical reading of the bibliography; conceptual analysis of the notion of autonomy, its normative bases and its technological substratum; and ethical-legal reflection on the risks of new technological developments when they affect aspects linked to personal autonomy.

3 Results

Of particular relevance to our work is Article 19 of the Convention, which establishes the right to live independently and to social inclusion. It refers to a variety of support mechanisms for the promotion of autonomy and the achievement of this right. Other articles of the Convention also refer to the protection of privacy, integrity and intimacy, given the fact that many technological devices that would have a place in Article 19 also have a place in these other articles.

Technological change has made the nature of assistive technologies for personal autonomy more complex. Many of today's technologies address the promotion of autonomy by monitoring behaviours, habitual behaviours and life patterns of people with disabilities. So-called *self-tracking data* technologies, using wearable technology and sensors, record behaviour and different physiological variables (vital signs, sleep patterns, electrical brain activity, etc.), allowing decisions to be made that empower individual autonomy or improve health. The data recorded are used to monitor the person and, thanks to digital devices and terminals, are processed by algorithms *in situ*, providing relevant information on health, habits, etc.

The massive collection of personal behavioural data did not occur in traditional technologies: technical aids or assistive products. The effects of new technological implementation may have repercussions on personal autonomy: think, for example, of how robotic health monitoring could invade privacy. Ethical reflection is therefore necessary to broaden the scope of the rights included in the Convention, thus protecting more effectively the autonomy of persons with disabilities. The changing technological scenario makes it necessary to take the protection of personal autonomy to the extreme, given the greater potential of these technologies to affect aspects related to it (privacy, intimacy, etc.). The rights of the Convention, suitably reinterpreted to take into account this potential, constitute an unavoidable guide in the ethical reflection on such technological risks.

The first step in the direction of a possible recognition of new human rights related to neuroscience is the debate on cognitive freedom. According to Bublitz [9]: Cognitive freedom, also sometimes referred to as mental self- determination, is the principle that guarantees the right to alter one's mental state through the use of neurotechnologies, as well as the right to refuse such action. This idea includes two closely related concepts: 1. The right of individuals to the use of emerging neurotechnologies, and 2. The protection of individuals against coercive use of these technologies, i.e. in case they are used without the consent of the persons concerned.

With regard to the second aspect, it may lead to the reconceptualisation of some existing rights, or to the creation of new fundamental "neuro-rights": the right to mental privacy, the right to mental integrity, or the right to psychological continuity.

4 Discussion

Technological change implies the redefinition of categories such as "personal autonomy" or "care". The reconfiguration of personal autonomy, brought about by technological changes that have a greater impact on rights such as privacy, security and data protection for people with disabilities, promotes new forms of care. It is no longer only about forms of personal care, but also about social care, i.e. the normative measures to preserve such rights, the same as those offered to any other person. Care must be interpreted as protection of one's own identity and individual conscience so that these new technological mechanisms do not become instruments of surveillance and control, curtailing the autonomy they were intended to promote.

These questions refer to the normative framework of the Convention. Let us turn, for example, to Articles 17 and 22 of the Convention, and others where these aspects are also addressed. There is no need to reinvent the wheel, but simply to get it rolling. To achieve this, it is necessary to take into account another dimension of personal autonomy: public autonomy [10]. This is embodied in the participation of persons with disabilities in the public debate on technologies, and in the development phases of technological artefacts, which is based on paragraph o) of the Preamble of the Convention, and is made concrete in several of its articles.

The importance of full participation in the technological design process on the part of users with disability must be emphasized. The users themselves know best how technological devices can contribute to their greater autonomy, quality of life and level of participation in society, and can assist in evaluating their advantages as well as their drawbacks in the earliest phases of their development [11]. There is no reason why technological development and innovation should remain solely in the hands of manufacturers and companies' R&D departments. Moreover, active user participation represents a source of innovation [12]. Some users, especially the so-called 'lead users' [13], conceive of possible advancements and improvements in the design phase that can be implemented and tested.

5 Conclusion

In this paper we try to show how reflection on change in the technological field of disability should motivate parallel reflection on the aspects of personal autonomy on which such technologies have an impact. The reflection presented here should not be overlooked, for if we do so, we will be leaving aside the consideration of a good number of rights set out in the Convention, with the consequent undermining of the consideration of persons with disabilities as citizens in their own right.

Acknowledgements. This contribution is presented within the projects INBOTS (Ref. 780073) and EXTEND (Ref. 779982) of the H2020 Programme, EthAI+3 (PID2019-104943RB-100), and the project "Capacitismo" (FFI2017-88787-R) of the Retos Programme of the Spanish State R+D+I Plan.

References

1. Honneth, A.: El derecho de la libertad. Esbozo de una eticidad democrática. Katz, Madrid (2014)
2. Winner, L.: Is there a right to shape technology? Argumentos Razón Técnica **10**, 305–328 (2007)
3. Nussbaum, M.: Las fronteras de la justicia. Paidós, Barcelona (2007)
4. Palacios, A.: El modelo social de la discapacidad: orígenes, caracterización y plasmación en la Convención Internacional sobre los Derechos de las Personas con Discapacidad. Ediciones Cinca, Madrid (2008)
5. Cayo, L., De Lorenzo, R. (eds.) La Convención Internacional sobre los Derechos de las Personas con Discapacidad – 2006–2016: una década de vigencia. Ediciones Cinca, Madrid (2016)
6. Convention on the Rights of Persons with Disabilities and Optional Protocol. UN, New York (2006)
7. Palacios, A., Romañach, J.: El modelo de la diversidad. La Bioética y los Derechos Humanos como herramientas para alcanzar la plena dignidad en la diversidad funcional. Ediciones Diversitas-AIES, Valencia (2006)
8. UN. Report of the Special Rapporteur on the rights of persons with disabilities (A/HRC/43/41). A thematic study on the impact of ableism in medical and scientific practice on persons with disabilities (2019)
9. Bublitz, J.-C.: My mind is mine!? Cognitive liberty as a legal concept. In: Hildt, E., Franke, A. (eds.) Cognitive Enhancement. Trends in Augmentation of Human Performance, vol. 1, pp. 233–264. Springer, Dordrecht (2013). https://doi.org/10.1007/978-94-007-6253-4_19
10. Habermas, J.: Facticidad y validez. Trotta, Madrid (2008)
11. Toboso, M.: Rethinking disability in Amartya Sen's approach: ICT and equality of opportunity. Ethics Inf. Technol. **13**(2), 107–118 (2011)
12. von Hippel, E.: The Sources of Innovation. Oxford University Press, New York (1988)
13. von Hippel, E.: Democratizing Innovation. MIT Press, Cambridge (2005)

Robotics, Childhood and Youth

Ana Lambea Rueda(✉)

Civil Law, Complutense University, Madrid, Spain
ana.lambea@der.ucm.es

Abstract. Robotics is one of the keys to the Technological Revolution. Its development is increasing and, to a certain extent, unknown.

Along with the benefits of robotics, there are important ethical and legal problems, especially in the interaction with the human beings in need of protection. As jurists, we must reflect on the robotics that we want to leave as a legacy to the next generations, today's children and young people.

In the European Union, in the field of robotics and artificial intelligence, there are matters of great interest, which should be explored to ensure the correct protection of children and youth. Thus, from a legal point of view, modifications of existing rules and/or new rules issued *ad hoc* are required.

1 Introduction

At the European level, there is a growing concern for citizens' rights in relation to artificial intelligence and robotics.

Indeed, the recent texts issued in this respect, show this in a rather generic way: the European Commission's White Paper on artificial intelligence[1] – a European approach to excellence and trust; and the European Parliament Resolution on a civil liability regime for artificial intelligence.[2] More detailed than the two previous ones is the Proposal for a Regulation on Artificial Intelligence[3], general as to its scope of application.

2 Present Situation

Some progress is being made towards an increasingly comprehensive regulation, the texts cited above refer to important issues in relation to minors and to robotics and

INBOTS CSA, "Inclusive Robotics for a Better Society", EU H2020 project, G.A. No. 780073.
AudIT-S, "Legal and financial significance of sustainability audit schemes through smart data management", Spanish National Research Plan, Ministry of Science and Innovation, PID2019-105959RB-I00.

[1] Brussels 19.2.020. COM (2020) 65 final.

[2] P9_TA (2020)0276. EP Resolution of 20 October 2020 with recommendations to the Commission on a civil liability regime for artificial intelligence (2020/2014 (INL)).

[3] Proposal for a Regulation of the European Parliament and of the Council laying down harmonizing rules on Artificial Intelligence (Artificial Intelligence Act) and amending certain legislative acts of the Union. COM/2021/206 final. COM/2021/206 final.

M. A. Grau Ruiz (Ed.): INBOTS 2021, BIOSYSROB 30, pp. 168–173, 2022.
https://doi.org/10.1007/978-3-031-04305-5_29

artificial intelligence. However, it is clear that they need to be further developed in the future, starting from a clear focus and direction.

With regard to Artificial Intelligence and people, the White Paper raises the protection of fundamental rights, the absence of discrimination, the areas of protection, without referring specifically or precisely to children and young people, minors.

In the area of civil liability, the European Parliament makes progress in the regulation of two regimes of strict liability and fault-based liability, depending on the seriousness of the risk, although the cases of high and low risk are not defined.

On the other hand, the Proposal for a Regulation covers the advanced issues in a fairly complete manner.

3 Texts Requiring Adaptation

Nowadays there are several texts that need to be adapted, either by means of specific modifications or other regulations that reflect their content.

With regard to the standards, we have noted and would like to highlight the following needs:

3.1 In General

The 2012 EU Charter of Fundamental Rights lacks references to the digital environment and robotics in relation to minors in Article 24.2. It could be incorporated.

The 1992 European Charter of the Rights of the Child does not include it either. It could do so, through one more addition, number 22, with reference to the protection of the child and his fundamental rights in this field, contemplating the best interests of the child as a principle of general and primordial application.

Intensive work has been done on Ethical Guidelines on Artificial Intelligence in the European Parliament, offered and recommended to the European Commission[4], although the two areas on which they focus are mainly data protection and education; which is, in my view, insufficient.

3.2 Specifically

With regard to minors, it would be of particular interest to amend Article 3.3.b of Regulation 2015/2120 of the European Parliament and the European Council of 25 November 2015 on open Internet. This would be to include as an exception to the open and universal Internet the action of minors in certain cases; and to require the offer of mechanisms for age control by Internet access providers and content providers.

The impact of the activity of minors in terms of data protection, included in Regulation 2016/679, is also fundamental. Thus, on the one hand, there is a need to review age issues, as well as specific mechanisms for the protection of minors when communication is through robots or AI.

[4] Report with recommendations to the Commission on a framework for the ethical aspects of artificial intelligence, robotics and related technologies (2020/2012 (INL)).

Thus, for the provision of consent, an age of 16 years may be proposed, and an absolute prohibition for minors under 7 years of age, adding restrictions or complements of capacity to consent for minors under 14 years of age.

It is important to distinguish several moments in the access and use of technology by minors, and the data concerned: before access (access, verification and restrictions); during use (geolocation and content); after disconnection (destruction, transmission or abandonment - necessary reformatting).

It is imperative to distinguish between: Consent for registration: registration, data; and actions as a user and rights affected by geolocation, content… It is also critical to highlight the possibility of alluding to the needs of minors in the E-Privacy Regulation.

On the other hand, the Directive 2018/1972 on Electronic Communications should take into account the needs of minors, to which it refers in its Article 96. It should include and develop the protection of minors, both in the Explanatory Memorandum and in Articles 2, 3 and 4.

In audio-visual communication material, the Services Directive, 2018/1808, would be affected in its recitals and in a large number of articles.

Likewise, it is also interesting to consider whether other rules require adaptation in the field in the case of interaction with minors. Examples include the Defective Products Directive 1985/374, as amended; the Toy Safety Directive 2009/48; the Copyright and Related Rights Directive 2019/790; the Information Society Services and Electronic Commerce Directive 2000/31; the Consumer Directive 1993/13; and the Misleading Advertising Directive 2006/114.

Some areas can be brought up that would also affect minors, in which their use should be prohibited, and are not expressly included, as is the case of autonomous vehicles; as well as the necessary precautions in assistive robotics for these groups.

The interaction of robotics and human beings requires the application of rules regarding the subject, the object and the content of the relationship. These rules are particularly important in the case of the most sensitive persons or users, such as minors, dependents, or the elderly. In the case of minors, it is necessary to distinguish according to age, depending on whether it is early childhood, puberty or adolescence; and to articulate a really effective protection in each case, with the provision of prohibition of absolute use in minors under 7 years of age; restricted in minors under 14 and with controls from 14 to 18 years of age. Of particular importance is age control by providers of access to robotics, and by providers of robotics-related content. With regard to training in robotics and artificial intelligence material for minors, the proposal could be: absolute exclusion in case of minors under 7 years of age, supervised training of minors from 8 to 14 years of age; and autonomous and adapted training of minors from 15 to 17 years of age. The distinction by age is fundamental, as well as the restriction and controls in each age group, even more so in some cases such as the acquisition of consumer goods or the use of autonomous vehicles.

It is necessary to insist on the need to incorporate in the European Charter of Children's Rights of 1992, article 22, that: "Every child has the right to be protected in his fundamental rights in the digital environment and Artificial Intelligence; contemplating the best interest of the child as a principle of general and primordial application".

4 New Standards?

A few years ago, it seemed clear that the need for new rules [raised by the European Parliament in successive Resolutions -2017, 2019, 2020-, in the 2018 European Commission Communication, and the 2018 Opinion of the European Economic and Social Committee]- was going to materialize quickly. Particularly pressing was the case of minors in the issues of protection, ethics, safety and responsibility in the fields of training, education, leisure, healthcare, assistance, autonomous vehicles, safer internet, network access, etc.

Today, after two years of global pandemic situation, there is no *ad hoc* standard, and the need for it is even greater given the exponential global spread of technology and robotics. The problems that were once thought to be possible are now obvious: data protection, ethics, security and liability. The previous section has presented some of the possible regulatory changes, such as the rules on fundamental rights in general and children's rights in particular, data protection regulations, open Internet, electronic privacy, some directives on advertising, toys, e-commerce, information society services... However, this has not been the line of action chosen by the European Commission, which seems to opt for new rules, specific to artificial intelligence and robotics, rather than modifying the existing ones.

Thus, the European Commission's Proposal for a Regulation on Artificial Intelligence of 21 April 2021 [COM (2021) 206 final 2021/0106 (COD)] can be cited. Reference should be made to it, although the debate may delay its effective entry into force and full implementation a few years. This proposal pays particular attention to the physical and psychological risks and harms that artificial intelligence can generate in natural persons. Some high-risk systems, related to human consumption, are specially identified with a specific list; and mandatory information to citizens of the interaction with artificial intelligence is required. With respect to this Proposal for a Regulation, doubts arise once again as to whether the appropriate formula is: whether an *ad hoc* regulation for artificial intelligence, or the amendment of the aforementioned rules.

Finally, the need for a future Regulation of the European Parliament and of the Council on ethical principles for the development, deployment and use of artificial intelligence, robotics and related technologies can also be mentioned. In particular, it should incorporate the protection of minors in specific articles, including the importance of protection of privacy and image rights; security and control of access and content; provision for physical and emotional risks and liability and penalties for non-compliance. Indeed, it is necessary to incorporate specific mechanisms for the protection of minors in the problems that affect them: awareness of the information they access, separation by age groups, protection technologies, verification of access and use, adoption of registration, licensing and insurance measures, and issues of identity and digital biography. Also, in the relationship of minors with artificial intelligence it is important to work on the image of the robot. The robot that interacts with minors cannot be an android robot with a human figure, as it may raise confusion, uncertainty and fear; especially in the younger ages.

5 Conclusion

It is necessary to reflect on the starting point of the interaction of robotics and artificial intelligence with human beings, especially those most in need of protection. A conscious review of the regulations that may be affected by robotics and artificial intelligence, in which minors, either in childhood or youth, may be involved and harmed, is proposed.

The reality of technology and robotics can be a useful means for the human being, and not an end in itself. It can support work and development, not restrict freedoms and enslave men. It can contribute to a better world, not support the destruction of the rights of humanity. The present moment is crucial, we are at a crossroads between two worlds. On the one hand, the world of the new normality, maintaining respect for the unique identity of the human being, his/her inviolable rights and freedoms. On the other hand, the robotized world in which robotics is an end in itself, with the risk of putting an end to the democratic rule of law which, since 1789, so many generations have defended with their lives.

References

1. Aparicio Torres, C., López Jara, M.: "La protección penal del menor víctima de ciberdelitos, Primeras actuaciones", Monográfico: menores y redes sociales. Rev. Ley Derecho Familia **14**, 51–71 (2017)
2. Barrio Andrés, M.: Del derecho de internet al derecho de los robots. Derecho de los robots, pp. 63–86. La Ley, Madrid (2018)
3. Brito Izquierdo, N.: "Tratamiento de los datos personales de menores de edad: supuestos, límites, retos y desafíos": Monográfico: menores y redes sociales. Rev. La Ley Derecho Familia **14**, 17–31 (2017)
4. Cazurro Barahona, V.: Menores de edad. La protección de su derecho a la intimidad en la era digital. Derecho Digital. Perspectiva interdisciplinar, pp. 243–276. Bosch, Barcelona (2017)
5. Díaz Alabart, S.: Robots y responsabilidad civil. Reus, Madrid (2018)
6. DíazDíaz, E.: El nuevo Reglamento General de Protección de Datos de la Unión Europea y sus consecuencias jurídicas para las instituciones. Rev. Aranzadi Doctrinal **6**, 1–32 (2016)
7. Guardiola Salmerón, M.: "Menores y nuevas tecnologías: los nuevos retos en el sector legal en España", Monográfico: menores y redes sociales. Rev. Ley Derecho Familia **14**, 3–16 (2017)
8. Lambea Rueda, A.: Robotics, cine y Derecho. Rev. BioDerecho **9**, 1–17 (2019)
9. Lambea Rueda, A.: Proteccion del menor en el entorno digital. In: López San Luis, R. (ed.) La protección del menor. Situation and Current Issues, pp. 177–207. Comares, Granada (2019)
10. Lambea Rueda, A.: Robotics and minors. In: Pons Rovira, J.L. (ed.) INBOTS 2018. Biosystems and Biorobotics, vol. 25, pp. 81–87. Springer, Cham (2019). https://doi.org/10.1007/978-3-030-24074-5_15
11. Lambea Rueda, A.: Entorno digital, robotica y menores. Rev. Derecho Ci, **4**, 183–232 (2018)
12. Lorente López, M.C.: Los derechos al honor, a la intimidad personal y familiar y a la propia imagen del menor. Thomson Reuters Aranzadi, Cizur Menor (2015)
13. Marañón Astolfi, M.: El interés del menor en los sistemas español y anglosajón. From the traditional divergence to the current convergence between legal systems. Rev. Derecho Privado **2**, 65–87 (2018)
14. Martínez Otero, J.M.: La protección jurídica de los menores en el entorno audiovisual. Thomson Reuters Aranzadi, Cizur Menor (2013)

15. Martinez Rojas, A.: Principales aspectos del consentimiento en el Reglamento General de Protección de Datos de la Unión Europea. Rev. Aranzadi Derecho Nuevas Tecnol. **42**, 59–82 (2016)
16. De La Cerda Ballesteros, J.A.M.: "La protección del honor de los menores en las redes sociales", Monográfico: menores y redes sociales. Rev. La Ley Derecho Familia **14**, 32–42 (2017)
17. Otaegui Azpurua, I.: La relevancia del Tribunal Europeo de Derechos Humanos en la protección de los derechos del menor. Thomson Reuters Aranzadi, Cizur Menor (2017)
18. Palmerini, E.: Robótica y derecho: sugerencias, confluencias, evoluciones en el marco de una investigación europea. Rev. Derecho Privado **32**, 53–97 (2017)
19. Paniza Fullana, A.: Una nueva era en la privacidad y las comunicaciones electrónicas: la Propuesta de Reglamento del Parlamento Europeo y del Consejo sobre el respeto de la vida privada la protección de los datos personales en el sector de las comunicaciones electrónicas. Rev. Doctrinal Aranzadi Civ.-Mercantil **7**, 1–12 (2017)
20. Ramón Fernández, F.: Menores y redes sociales: cuestiones legales. Rev. Sobre Infancia Adolescencia **8**, 33–44 (2015)
21. De Huidobro De Carlos, J.M.R.: La capacidad de obrar y la responsabilidad de los menores. In: Martínez García, C. (ed.) Tratado del menor. La protección jurídica de la infancia y la adolescencia, pp. 157–219. Thomson Reuters Aranzadi, Cizur Menor (2016)
22. Sánchez García, A.: Robotica y ética. In: Barrio Andres, M. (ed.) Robot Law, pp. 229–267. La Ley, Madrid (2018)
23. De Castro, E.V.: Protección de datos personales, redes sociales y menores. Rev. Aranzadi Derecho Nuevas Tecnol. 1–30 (2012)

Check for updates

Interactive Robotics and Children's Human Rights

Manuel Aparicio[1](✉), Mario Toboso[2], Ricardo Morte[3], Aníbal Monasterio Astobiza[3], Txetxu Ausín[4], and Daniel López[4]

[1] Faculty of Philosophy, University of Murcia (UMU), Murcia, Spain
manuel.aparicio@um.es
[2] Institute of Philosophy, Spanish National Research Council (CSIC), Madrid, Spain
mario.toboso@csic.es
[3] Faculty of Philosophy, University of Granada (URG), Granada, Spain
ricardo63@autistici.org, amastobiza@ugr.es
[4] Institute of Philosophy, Spanish National Research Council (CSIC), Madrid, Spain
{txetxu.ausin,daniel.lopez-castro}@cchs.csic.es

Abstract. Developments in AI and robotics in recent years have opened new possibilities for the introduction of these technological devices in areas such as care, education, and entertainment, with important implications for children. The rise of interactive robotics in these fields requires that technological designs involve the impact that such interaction may have on children, considering the special needs for the protection of dignity and the growth of autonomy that characterises this stage of life. In this paper we argue that this valuing perspective on childhood introduces the requirement that robotic designs and their uses must normatively conform to the human rights of children.

1 Three Domains of Robot-Child Interaction

Developments in interactive robotics have three important areas of application in care, entertainment and education, all of can involve children. However, robots are not merely instruments to serve certain purposes related to these activities but can also lead to a transformation of what we understand as care, play or education [5]. On the other hand, robots may not always contribute to the desirable expansion of children's capacities; they may also contribute to their diminution [12]. It is therefore urgent to reflect on the limits of introducing this new technology [19]: what uses of these artefacts may be appropriate in childhood, so that their mediation could help to promote balanced development at this stage of growth for all human beings.

In relation to the three areas mentioned above, certain risks can be identified that need to be addressed:

This contribution is presented within the projects INBOTS (Ref. 780073) and EXTEND (Ref. 779982) of the H2020 Programme, EthAI+3 (PID2019-104943RB-100), and the project "Capacitismo" (FFI2017-88787-R) of the Retos Programme of the Spanish State R+D+I Plan.

M. A. Grau Ruiz (Ed.): INBOTS 2021, BIOSYSROB 30, pp. 174–178, 2022.
https://doi.org/10.1007/978-3-031-04305-5_30

1. Consider the protective and empowering importance of care for children, contributing to the life-sustaining, physical health, mental well-being, and emotional development of children. Various works point out the risks of the substitute use of robots in care tasks, more accentuated in repeated long-term care, given the negative impact on the cognitive, emotional, and social development of the child, with repercussions in their adult life [10, 13, 16], achieving a balanced development of the different facets of the personality limits technological care, as it becomes dehumanising and neglectful care. On the contrary, care that facilitates the balanced development of children requires the complementary use of interactive robots, under the supervision of parents or other caregivers, considering the different degree of autonomy achieved in the various stages of childhood [2]. In support of human care, both in domestic and institutional contexts (hospitals, schools, etc.) interactive robots can contribute to protecting children's basic rights such as life, health, safety, and bodily integrity.
2. Many countries have introduced "educational robotics" at different stages of education, aimed at the design, construction, and programming of robots. This technological education helps children and adolescents to be in a better position to understand and to be able to cope vitally with an increasingly robotised world. On the other hand, the introduction of robots in the school world needs to be carefully considered in terms of their contribution to the right to education, because of the repercussions that the replacement of teachers can have on children's emotional development [4]. Other risks derived from the use of robots could be the decrease of social skills [11], techno-brain exhaustion due to the abuse of technology [15], or a probable weakening of some cognitive skills when used as 'slave' tools [3] in learning tasks. A complementary use of educational robots can be motivating and thus facilitate autonomous learning of certain content. An education complemented with robots, promoting scientific- technological literacy and a humanistic education that reflects on the meaning and limits of technology, will prepare students to live in an adult context of appropriate interaction with robots.
3. Play has a constitutive and valuable character for the dignity of children. Play time and the materials used in play should contribute to the harmonious development of the child's personality. Robots for entertainment, in accordance with age and growing capacities, could contribute to the harmonious development of the child, but should be safe from the point of view of health and physical and mental integrity. Depending on their type, they should be designed to support the development of imagination and thinking, emotions, social relations, physical development, but they should be designed to avoid discriminatory biases. It should be noted that the type of entertainment robot and the contents that robotic devices may include must be appropriate to the age of the children. In the design of robots for adolescents, in anticipation of their use to harm others, cybersecurity measures should be introduced, as well as reinforced measures of legal responsibility and civic- moral education. As for sex robots, the risk of their use lies in the possible objectification of the sexual partner, detached from an affective relationship, or in access at an inappropriate age. Parental control and appropriate affective- sexual training contribute to the balanced development of this human dimension, mitigating this technological risk.

2 Dignity and Human Rights in Child-Robot Interaction

The responsible incorporation of interactive robotics in activities (care, play, education) in which children are socially involved requires the existence of standards that develop the ethical-legal requirements of protection, full and balanced development and well-being underlying the Convention on the Rights of the Children (CRC) [17]. It should be noted that the current understanding of childhood [6, 14] considers it as a vital stage subject to a higher vulnerability, which demands special protection, and in which there is a progressive development of physical, psychological, and moral capacities that are at the basis of a continuously growing autonomy. From this perspective, childhood is a period of life that requires normative support for its own sake and for the consequences it will have on adult life. Such normativity is justified, as expressly reflected in the Preamble of the CRC, by the recognition of "the inherent dignity and the equal and inalienable rights of all members of the human family". By virtue of the existence of a conceptual connection between dignity and human rights [1].

[1] the dignity that is also recognised for children implies recognition of their human rights. It is therefore congruent to protect the vulnerability and promote the development of children's autonomy also in the new socio-technical context of child-robot interaction.

Other relevant aspects of the CRC to be considered are the concept of the best interests of the child, which implies that the legal rules regulating care, educational or play robotics must involve the general interests of children. Or article 12 of the CRC, which obliges, based on maturational development, that children must be listened to in the design and uses of robots that are to interact with them.

It is therefore imperative not to forget that the epistemic and technical aspects underlying the development of robotic technology must be coordinated with the normativity derived from this international legal-moral document.

3 Ethical Requirements for Robotics Design

The dual moral and legal nature of human rights [9] not only requires national and international legal regulation to frame the design and use of this technology; it also implies that robotics for interaction with children adopt an ethical approach that protects their vulnerability and promotes the balanced and harmonious growth of their autonomy. Those involved in the design and physical construction of robotic devices, their virtual programming or responsibility for their uses need to cope with a set of guiding ethical principles [7, 8] that fill the gaps or deepen what is established in the legal norms:

Non-maleficence principle: robotic designs have to be subjected to rigorous testing processes to ensure, within demanding parameters, physical and emotional safety and integrity, data protection, the right to privacy, cybersecurity or the avoidance of discriminatory bias in design or programming.

Beneficence principle: robotic designs and uses of robotic devices for children should contribute to the improvement of their quality of life, being additional facilitators of balanced human development. To this end, the design and use of artefacts in the interaction with children should be adjusted to the dynamic nature of child development, taking into account the evolution of functionings and capacities over the time included in this life stage.

Principle of autonomy: the design and uses of child- directed robots should facilitate the balanced development of the autonomy-generating dimensions of the personality. They should also be designed to retain the ability to decide which tasks are delegated to robots [8] or even to switch them off when danger is foreseen. Moreover, children with functional diversity should be recognized, in accordance with the principle of universal accessibility as set out in the Convention on the Rights of Persons with Disabilities [18], so that inclusive robotics also takes technological care of their autonomy and thus facilitate their active participation in society. This requires that robotic designs incorporate accessibility features, contribute to facilitating accessibility for children with functional diversity in different environments, and are compatible with the actual features of those environments.

Principle of justice: equitable access to robotic technologies must be promoted to avoid social gaps. On the other hand, it also follows from this principle that the design, programming, and use of robots aimed at children should avoid the existence of any discriminatory bias.

Principle of explainability: related to understanding how it works, accountability and responsibility.

It is therefore necessary for those involved in the development of interactive robotics technology to have a solid ethical background, complementary to their technical training. Comprehensive training, the codes of ethics of the professions involved, and corporate responsibility will help to reinforce confidence in the social appropriation of interactive robotics for children.

4 Conclusions

The predominant hermeneutic perspective of a value- based understanding of childhood as a stage of equal dignity, characterised by greater needs for protection and growth of autonomy that should be fairly promoted, constitutes an inescapable reference in the introduction of child-interaction robotics. The protection of their dignity through the set of their human rights, already internationally committed to in the CRC, taking into account, as a matter of priority, the best interests of children and the obligation to listen to them, in accordance with their growing autonomy, in matters affecting them, introduces a set of legal requirements and refers to a set of guiding ethical principles in the new developments of interactive robotics. Regulatory legal standards, compatible with children's human rights, and ethical training and enforcement in engineering and business culture will reinforce trust in such artefacts, increasing the social acceptability of child-robot interaction. These normative conditions constitute both a protective boundary and a source of promotion for the just human development of children. With them, robotic designs introduced into the care, education and entertainment of children will better contribute to these goals. It is up to the groups interested in this technology to make them their own.

References

1. Alexy, R.: La doble naturaleza del derecho Madrid. Trotta (2016)
2. Biggeri, M., Karkara, R.: Transforming children's rights into real freedom: a dialogue between children's rights and the capability approach from a life cycle perspective. In: Stoecklin, D., Bonvin, J.-M. (eds.) Children's Rights and the Capability Approach. Challenges and Prospects, vol. 8, pp. 19–41. Springer, Dordrecht (2014). https://doi.org/10.1007/978-94-017-9091-8_2
3. Carr, N.: Atrapados. Cómo las máquinas se apoderan de nuestras vidas. Taurus, Madrid (2014)
4. COMEST: Informe de la comisión sobre la ética robótica. UNESCO, París (2017)
5. Coeckelberg, M.: "How i learned to love the robot": capabilities, information technologies, and elderly care. In: Oosterlaken, I., van den Hoven, J. (eds.) The Capability Approach, Technology and Design. Philosophy of Engineering and Technology, vol. 5, pp. 77–86. Springer, Dordrecht (2012). https://doi.org/10.1007/978-94-007-3879-9_5
6. Drerup, J., Graf, G., Schickhardt, C., Schweiger, G. (eds.): Justice Education and the Politics of Childhood. Challenges and Perspectives. Springer, Switzerland (2016)
7. EGE (European Group on Ethics in Science and New Technologies): Statement on Artificial Intelligence, Robotics and 'Autonomous' Systems. Publications Office of the European Union, Luxembourg (2018)
8. Floridi, L., et al.: AI4People – an ethical framework for a good ai society: opportunities, risks, principles, and recommendations. Atomium -European Institute for Science, Media and Democracy, Brussels (2018)
9. Habermas, J.: El concepto de dignidad y la utopía realista de los derechos humanos. In: Habermas, J. (ed.) La constitución de Europa. Trotta, Madrid (2012)
10. Hosseini, S.H., Goher, K.M.: Personal care robots for children: state of the art. Asian Soc. Sci. **13**(1), 169–176 (2017)
11. López, A.M., Castro, A.: Adolescencia. Alianza editorial, Madrid (2007)
12. Oosterlaken, I.: The capability approach, technology and design: taking stock and looking ahead. In: Oosterlaken, I., van den Hoven, J. (eds.) The Capability Approach, Technology and Design. Philosophy of Engineering and Technology, vol. 5, pp. 3–26. Springer, Dordrecht (2012). https://doi.org/10.1007/978-94-007-3879-9_1
13. Sharkey, A., Sharkey, N.: Children, the elderly, and interactive robots. IEEE Robot. Autom. Mag. **18**(1), 32–38 (2011)
14. Shaw, M., Bailey, S. (eds.): Justice for Children and Families. A Developmental Perspective. Cambridge University Press, Cambrige (2018)
15. Small, G., Vorgan, G.: El cerebro digital. Como las nuevas tecnologías están cambiando nuestra mente. Urano, Barcelona (2008)
16. Torras, C.: Assistive Robotics: Research Challenges and Ethics Education Initiatives. Dilemata Rev. Int. Éticas Aplicadas **30**, 63–77 (2019)
17. UN: Convention on the Rights of the Child (1989)
18. UN: Convention on the Rights of Persons with Disabilities (2006)
19. Winner, L.: La ballena y el reactor. Gedisa, Barcelona (2008)

Analysis of Sustainability Reports in Two Industrial Companies

E. García Jara[1](✉), J. Muñoz Jiménez[2], and A. Prado Martín[1]

[1] Financial Administration and Accounting, Complutense University of Madrid (UCM), Madrid, Spain
{elisagj,aprado}@ccee.ucm.es

[2] Financial Administration and Accounting, Rey Juan Carlos University (URJC), Madrid, Spain
jose.munoz@urjc.es

Abstract. The objective of this paper is to present, through the analysis of the sustainability reports presented by two representative companies of the robotics sector, their commitment to Corporate Social Responsibility. This work is the germ of a subsequent research with a wider scope. As a result of this analysis, we can see that the commitment of these companies to the fundamental principles of CSR, such as its recognition and the involvement of stakeholders, is increasing, meeting the objectives in the various key areas.

1 Introduction

"Sustainable development consists of meeting the needs of society while respecting the ecological limits of the planet and without compromising the ability of future generations to meet their needs".

Sustainable development has three dimensions: economic, social and environmental, which are interdependent. It can be used as a way of summarizing society's broadest expectations that need to be taken into account by organizations seeking to act responsibly. Therefore, the primary objective of a socially responsible organization should be to contribute to sustainable development.

Social responsibility has the organization at its center of interest and concerns an organization's responsibilities towards society and the environment.

To define the scope of its social responsibility, identify relevant issues and establish its priorities an organization should address the following key issues:

- governance of the organization;
- human rights;
- labor practices;
- environment;
- fair operating practices;
- consumer issues, and
- active participation and community development" [1].

The Corporate Social Responsibility (CSR hereafter) reports of the companies: Ottobock for the fiscal year 2019 and Össur for the fiscal year 2020 have been analyzed.

M. A. Grau Ruiz (Ed.): INBOTS 2021, BIOSYSROB 30, pp. 179–184, 2022.
https://doi.org/10.1007/978-3-031-04305-5_27

2 Analysis of the Ottobock Company Report Leadership of the Electronic Tax Agency

"It is a German company founded in 1919 by Otto Bock, in 2019, its 100th anniversary, is 80% owned by Näder Holding GmbH & Co. KG wholly owned by the owner family, the Näders" [2], and "by the EQT fund, which is one of the big names in the European private equity industry. Controlled by the Wallenbergs, one of Sweden's most influential families" [3].

Its most significant figures in 2019 were:

- EUR 1,003 million turnover in 2019.
- More than 7,000 employees in more than 50 countries worldwide.
- More than 160 patient care centers operated by the company.
- World leader in prosthetics, especially knee prosthetics.

The areas that form the basis of its CSR activities:

1. patients/users
2. Product quality
3. Medical personnel
4. Innovations
5. Materials
6. Partnerships
7. Human capital
8. Fair business practices/compliance

Being committed to the 10 principles of the UN Global Compact for sustainable development which are:

Human Rights:

Principle 1: Businesses should support and respect the protection of internationally proclaimed human rights.
Principle 2: Businesses should make sure that they are not complicit in human rights abuses.

Labor:

Principle 3: Businesses should uphold the freedom of association and the effective recognition of the right to collective bargaining.
Principle 4: Businesses should uphold the elimination of all forms of forced and compulsory labor.
Principle 5: Businesses should uphold the effective abolition of child labor.

Principle 6: Businesses should uphold the elimination of discrimination in respect of employment and occupation.

Environment

Principle 7: Businesses should support the elimination of discrimination in respect of employment and occupation.
Principle 7: Businesses should support a precautionary approach to environmental challenges.
Principle 8: Businesses should undertake initiatives to promote greater environmental responsibility.
Principle 9: Businesses should promote the development and diffusion of environmentally friendly technologies.

Anti-corruption

Principle 10: Businesses should work against corruption in all its forms, including extortion and bribery [4].

3 Össur Company Report Analysis

It is an Icelandic company based in Reykjavik and founded in 1971 by Össur Kristinsson. It is currently a world leader in non-invasive orthopedics, specifically innovating, producing and marketing prostheses as well as orthoses and supports. Listed on the stock exchange since 1999 (NASDAQ Copenhagen), it is present in 27 countries and employs more than 3,500 people [5]. The economic, financial and results data for the first quarter of 2021 (consolidated data in US dollars) are as follows: sales amounted to 160.8 million, with a net profit of 11.4 million; R&D expenses reached 7.3 million; assets amounted to 1,213.2 million; liabilities were 629.5 million; and with a net worth of 583.7 million [6].

The company contributes fundamentally to four of the UN Sustainable Development Goals:

1. Goal 3: Good health and well-being.
2. Goal 5: Gender equality.
3. Goal 12: Responsible consumption.
4. Goal 13: Climate action.

Össur contributes mainly to Goal 3 on good health and well-being by producing and marketing, as its products improve and give people access to appropriate solutions with their prostheses, orthoses and supports. In particular, they provide solutions for amputations in the elderly. 59% of its sales are mechanical and biomechanical prostheses (intended for amputations), the rest are sales of orthoses and supports (intended for preventive and therapeutic purposes to recover from fractures, injuries, etc.) The company has R&D expenses of 4.53% of sales (first four months of 2021). In its CSR report it indicates an average of 5%, which is in line with the information provided in the report. In the 2020 financial year, the company has launched more than 20 new products on the market, with great development in innovation as a result of R&D expenditure.

In relation to objective 5 on gender equality, the company has 54% men compared to 46% women, meeting the proposed objective. In addition, women occupy 36% of jobs in leadership and management positions. Össur has both internal and external Women's Leadership programs to meet the needs of women in the orthopedics and prosthetics sector and thus achieve the proposed objectives.

Responsible consumption as the 5th objective, the company has a responsible sourcing process with regular audits.

In relation to the 13th objective on climate action, the company aims to minimize the environmental footprint, for this purpose it has several programs. In addition, it has a recycling rate of 65%.

The private companies analysed are highly committed to sustainable development, as evidenced by their compliance with sustainability standards.

The requirement for reliable and transparent information contained in non-financial statements on environmental issues is also a concern of public bodies, expressed by regulators and supervisors of the information issued by companies, due to their impact on sustainable development.

As Mr. Santiago Durán, President of the Instituto de Contabilidad y Auditoría de Cuentas (ICAC) indicated in his presentation at the UCM Seminar entitled "Audit of Annual Accounts and Non-Financial Information Statements: Contributions to Sustainability", sustainable development is based on the balance of three elements: economy, society and environment. The aim is to ensure that economic progress reaches everyone, that equality is achieved, and that the environment is protected.

The importance of the reliability of the information published by companies will play a fundamental role in a situation emerging from an economic crisis, in which a recovery plan involving transformation and a set of structural reforms must be tackled.

The transparency and reliability of non-financial information is a control mechanism over the company itself for this recovery.

This requires a regulatory framework, which in the case of Spain is regulated by Law 11/2018, which amends the Commercial Code and the revised text of the Spanish Capital Companies Act and is the result of the transposition of the 2014 Directive on disclosure of non-financial information or diversity information by large companies and the consolidated text of the Capital Companies Act and is the result of the transposition of the 2014 Directive on disclosure of non-financial information or information on diversity by large companies. Identifying risks to increase sustainability and gain the trust of all users of this information.

Following the aforementioned Law 11/2018 in Spain, the ICAC has published an informative GUIDE on the application of Law 11/2018, of 28 December, which amends the Commercial Code, the revised text of the Capital Companies Act approved by Royal Legislative Decree 1/2010, of 2 July, and Law 22/2015, of 20 July, on auditing of accounts, in terms of non-financial information and diversity.

The National Securities Market Commission (CNMV) is responsible for supervising the non-financial information of listed companies with more than 500 employees during the financial year, which in 2021 has been reduced to 250.

The Non-Financial Information Statements have become a very important document forming part of the Management Report and included in the annual report that these companies must issue.

At the international level, the European Securities and Markets Authority (ESMA), the body that brings together all the securities commissions of the European Union, is the one that establishes the information priorities required from the different national bodies. In recent years ESMA has considered the reporting of non-financial information statements to be a priority.

According to information provided by Ms Elena Peña Molina (CNM representative) at the UCM Seminar (2021), 70% of the groups that submit consolidated financial statements have submitted the required non-financial information statements. It should be noted that, in the last review, there were no qualifications, which shows the involvement of the companies in improving and adapting to these new requirements.

At present, the degree of maturity of the financial information with respect to the non-financial information could be compared through the following variables:

- Responsibility, in this case we can say both for the issuance of financial and non-financial information the responsibility lies with the Board of Directors of the company, therefore in this case the degree of maturity reached would be 100%.
- Formal corporate policies exist for both financial and non-financial reporting in IBEX 35 companies. Therefore, the degree of maturity also reaches 100%.
- Company strategies, in non-financial reporting, are still not applied in more than 50% of IBEX companies.
- Relevance of information, this is mature in financial information. However, in the field of sustainable reporting it is mature in view of the standards, but not yet in reality. It is a difficult concept to develop.
- Financial risk management is mature. In non-financial reporting sustainability risks are not yet mature.
- Verification is ensured in financial reporting. However, given its complexity, as it covers many areas, only some IBEX 35 companies are implementing verification systems for non-financial information.

In short, a sufficient degree of maturity has not yet been reached to have a single way of calculating sustainability information.

At the European level, work is also being done in this direction. The European Financial Reporting Advisory Group (EFRAG) has submitted two reports, in February and March 2021, to the European Commission with recommendations on the scope and structure of non-financial reporting statements and on EFRAG's own structure to ensure that future EU sustainability reporting standards are rigorously developed.

Sustainability reporting aims to provide relevant, accurate, comparable, and reliable information on the environmental impacts of the entity, the material sustainability risks it may produce for its own value creation.

The European Commission has presented a package of measures to help improve cash flow towards sustainable activities across the European Union in line with the European Green Pact and the UN Sustainable Development Goals.

This package includes a proposal for a Directive on non-financial reporting by companies on sustainability that would amend Directive 2013/34/EU (as amended by Directive 2014/95/EU of the European Parliament and of the Council of 22 October 2014) Directive 2004/109/EC, Directive 2006/43/EC and Regulation (EU) No 537/2014, as regards sustainability reporting by companies.

The objectives are:

Create a set of standards that align sustainability reporting with financial reporting.

- Ensure that companies report reliable and comparable sustainability information needed by investors and other information users.
- Simplify the reporting process.

It is expected that by 2024 companies would apply the standards for the first time in 2023 reporting.

4 Conclusion

Based on the reports submitted by the companies studied we can conclude that:Both companies are committed to the 10 principles of the UN Global Compact for sustainable development, as indicated in their CSR report and reinforced by their implemented programs.

References

1. UNE- ISO Standard 26000 Guide to Social Responsibility
2. The Ottobock company: Corporate Social Responsibility 2019
3. https://www.eleconomista.es/empresas-finanzas/noticias/10764392/09/20/Quien-es-EQT-el-nuevo-dueno-de-Idealista-y-que-planes-tiene-con-el-portal-de-vivienda.html
4. https://www.un.org/es/cr%C3%B3nica-onu/el-pacto-mundial-de-la-onu-la-b%C3%BAsqueda-de-soluciones-para-retos-globales
5. The Össur compay: Corporate Social Responsibility 2020
6. First quarter 2021 data from Óssur. https://media.ossur.com/image/upload/v1619473466/documents/corporate/Quarterly%20Reports/2021/2021_Q1_%C3%96ssur_hf_Financial_statement.pdf. Accessed 10 May 2021
7. Bryan, K., Castilla, A., Durán, S., Peña, E.: Auditing of annual accounts and the statement of non-financial information: contributions to sustainability. In: Prado, A. (ed.) Academic Events of the Faculty of CC. Economics and Business Administration. UCM, Madrid, 15 December 2021. https://economicasyempresariales.ucm.es/buscador/?bci=videos&search=Auditor%C3%ADa%20de%20Cuentas%20Anuales%20y%20del%20Estado%20de%20Informaci%C3%B3n%20no%20Financiera:%20Aportaciones%20a%20la%20Sostenibilidad

Intelligent Management and Control of Data Contained in Non-financial Reporting

A Proposal on the Applicability of Robotics, Artificial Intelligence and Blockchain

Virginia Martínez Torres(✉)

Financial and Tax Law, Complutense University of Madrid, Madrid, Spain
virgim05@ucm.es

Abstract. The Law calls on technology so that robotics, Artificial Intelligence and the Internet of Things can provide consistent information, analyze the temporal evolution of companies in terms of sustainability and, supported by systems that guarantee the origin and traceability of data, harmonize non-financial information to achieve the same level of trust, transparency and commitment as that provided by Financial Reporting Statements.

1 Introduction

Environmental law is proving to be one of the most evolving branches of law as climate change is one of the most pressing concerns on today's political agendas. The regulatory development of Non-Financial Reporting (NFR) must be closely linked to developments in sustainability, environment, climate change and circular economy. However, sometimes rapid legislation leads to shortcomings in control systems, resulting in a lack of effectiveness and efficiency.

The mission of these lines is to propose a system of management and intelligent control of the data in the NFR that will improve the quality of the information and the rigorousness of the controls. In addition, we believe that, when properly connected to environmental taxation, tax law and its capacity to stimulate certain behaviors in society will be able to achieve the real objective: the reduction of negative externalities.

2 The Control Systems Required in the NFR by the European Union at the Present Time

In 2014 the European Union published Directive 2014/95/EU amending Directive 2013/34/EU regarding the disclosure of and diversity information by certain large undertakings and groups [1]. This amendment entails the inclusion of the new Non-Financial

AudIT-S, "Legal and financial significance of sustainability audit schemes through smart data management", Spanish National Research Plan, Ministry of Science and Innovation, PID2019-105959RB-I00.

M. A. Grau Ruiz (Ed.): INBOTS 2021, BIOSYSROB 30, pp. 185–191, 2022.
https://doi.org/10.1007/978-3-031-04305-5_31

Reporting, thus consolidating the obligation for certain companies to materialize what in Directive 2013/34/EU [2] remained under voluntary nature.

The justification for these modifications focused mainly on the need to increase the transparency and comparability of social and environmental information. It is true that transparency has played a central role and has been able to materialize and publicize the NFR, but no care has been taken to preserve the chain of custody of this information, nor to guarantee its veracity through adequate internal and external controls. On the other hand, the objective of achieving the desired comparability of this information has not been achieved either, due to its high degree of heterogeneity and the lack of common frameworks and standards.

The controls that Directive 2014/95/EU foresees on NFR are based on a simple verification of having formulated such NFR by the auditors and, open the possibility for the States to incorporate in their internal rules a review by a verifier to ensure that the report has been prepared in accordance with the regulations, i.e., a limited verification that will not guarantee that the data contained in the NFR are true.

In 2021, the Proposal for a Directive on corporate reporting on sustainability [3], raises the requirement for this limited verification in the NFR. This limited verification (reviews) will confirm that no evidence has been obtained to conclude that the reporting has not been fairly presented in all material respects, so that the assurance they provide is acceptable and their cost is considerably lower. These procedures are more limited in comparison with those desired, which provide reasonable assurance (audit) and lead to a conclusion in the affirmative, i.e., that the matter under review complies with all significant aspects set out in the standard and represents a true and fair view, free from material misstatement and provides an acceptably low level of risk.

The enunciation of purposes, such as transparency and comparability, devoid of means of control cannot be qualified as initiatives aimed at adaptation or mitigation of externalities. Even if these objectives had been achieved, the legislator did not have sufficient foresight to make this system of obligations a useful system. It has resulted in high costs for companies and for the administration without any results in favor of sustainability, since no rules have been envisaged to guide the behavior of companies to improve their performance.

However, it is true that it has been possible to implement a system in which, for the time being, the entities subject to the standard comply with their obligations. They have been imbued, to a greater or lesser extent, with a system that extols the importance of sustainability in society and has familiarized them with certain obligations that, if well managed, can bring significant social, economic and environmental benefits.

From our point of view, we believe that the right thing to do is to give priority to objectives such as contributing to the transformation of the economic system, making it climate neutral and circular, without losing sight of others that we understand to be secondary, such as transparency and comparability, which will undoubtedly help to reinforce investor security, improve corporate governance and guarantee correct information to consumers.

3 The Possibilities of Improving NFR Controls Through Robotics, Artificial Intelligence and Blockchain

Following on from the above, it should be noted that the internal control of non-financial information, which is less mature than the control of financial information, should consist of an organizational plan and a set of protocols to ensure that the records containing non-financial information are reliable, accurate and complete. Their lesser degree of development means that external controls are of less practical use, and this is due to the fact that companies do not currently have internal control mechanisms to ensure the accuracy of the data collected. In addition, the reporting systems of the issuing entities, as they are not standardized or automated, often do not allow for the adequate collection of information.

Law and technology do not share the same pace, as technological development is often ahead of what is expected from law, far surpassing it. Generally speaking, the Fourth Industrial Revolution projects a context in which it is necessary to adapt legal rules so that they can adapt to the new reality. However, as we shall see below, on this occasion the law will demand an update from technology to help realize its expectations.

Therefore, we believe that the regulation of internal controls and the creation of robots, artificial intelligence systems and the implementation of the internet of things, which combined will provide quality and security to the development of this process and, finally, will result in greater efficiency and effectiveness in external controls, are advisable.

We advance the global idea of this proposal which consists, in broad strokes, in the creation of an Artificial Intelligence that is capable of measuring the externalities produced by the obligated entities and, connected to a private blockchain, guarantees the reliability of the registration of this information. In addition, this Artificial Intelligence will propose a model of the NFR and, will improve through the blockchain registry the learning capacity of the system and will decrease human error. The work of the internal auditors will be the commentary and justification of the NFR proposed by this Artificial Intelligence. On the other hand, the independent verifiers, in charge of the external control, will be in charge of verifying that the information is compatible and assessing the comments and justifications of the former, to approve the data that will finally contain the NFR.

In the following paragraphs, we will focus on developing the operation of this Artificial Intelligent (AI) and the advantages of the private blockchain as a method of recording Non-Financial Reporting.

For the development of these premises we must start from the use of new technologies associated with the internet of things, which allows us to install Artificial Intelligence in existing infrastructures through intelligent devices that can communicate. This could be through interactive robots equipped with artificial intelligence, i.e., intelligent devices with the ability to communicate, strategically located according to the externality. They would be designed to measure and report their status and could develop through the information that the entities provide on their activities an overall estimate data on their impact on the environment. Another option would be to start from the creation of these instruments and evaluate how to implement them in the elements causing the externalities, in order to make these elements capable of capturing the necessary information to measure the activity and the externalities, so that they collect the set of data required for the development of useful indexes.

For example, autonomous vehicles already provide assistance and organization services, such as choosing the most suitable routes and schedules. These systems that are already implemented in this type of vehicles could be expanded and improved to allow centralizing the information of the entire fleet of vehicles of an entity in such a way that they would be able to generate a report where all the necessary data for the computation of the negative externalities emitted by these vehicles would be collected. This approach would involve a combination of robotics, Artificial Intelligence and internet of things in different modalities and adjusted to the specific case. The idea would be to replicate this idea in all those main sources of externalities such as: smoke from factories in terms of greenhouse gas emissions, the different sources of CO_2 emissions or waste discharges, among others.

The use of private blockchain systems will allow the automatic aggregate valuation of the relevant data of all the subsidiaries of the entity subject to the preparation of the NFR. The European Directive exempts subsidiaries from the submission of NFR as long as their information is captured by the report submitted by the parent company. This data automation system would mean a significant saving of time and effort that these entities currently invest in gathering all this information. This will be possible through a mesh network configuration, consisting of a network that connects devices with each other. Another advantage would be the possibility for these devices to automatically reconfigure themselves, making them resistant to breaks or other forms of interruption, or to warn of excessive externalities at certain points so that they can be corrected. Today, the combination of the mesh network with the private blockchain is the solution to solve complex infrastructure problems and to achieve greater security and privacy.

This data will be recorded in a private blockchain. The main feature that differentiates private blockchain from public ones is its limited access, as it restricts the people who can use the network and this leads to a series of differences in its program that particularly benefits companies that wish to use them, in fact, they are born from the need of the private sector to use them. Currently, the private blockchain is being valued mostly by industry for use as a distributed ledger, as it allows entities to implement a traceable, immutable and secure storage system.

In our proposal, the Artificial Intelligence will make records in the private blockchain system, which will be built under the "Execute-Order-Validate" architecture, through which the Artificial Intelligence will act as an executor node called endorser with the capacity to make reliable executions in the network. This can be so, thanks to the fact that the private blockchain dispenses totally or partially with the "round-trip" verification systems required by the public blockchain. It allows the interaction between data to be more efficient, requires less processing capacity and, energy costs are significantly reduced.

Once the information has been collected and when certain conditions occur, the response will be sent to the ordering service, which orders the transactions, adds the block to the chain and sends it to the network. In other words, a node with rights to Artificial Intelligence will be enabled to ensure the correct registration of the information.

The private blockchain can constitute for NFR a new type of database that certifies the production of such externalities, guaranteeing that each transaction entered by Artificial Intelligence is true and correct and preventing such information from being manipulated. In addition, this registration system will facilitate the auditing of the data from its creation, guaranteeing its transparency, traceability and immutability. At this point, we are providing, at the same time, security to the information contained in the NFR and guaranteeing a correct auditability.

The verification of these records and verification of the NFR proposed by Artificial Intelligence will be the responsibility of the internal auditors, who will ensure that this information is presented in a way that guarantees a true and fair view of the company, complying with the guarantees of reliability. The internal auditor will be provided with a node with read permissions that will also be able to make proposals for improvement. Some sort of automated control for certain details could also be envisaged.

Subsequently, the independent verifiers within the external control will have another node with extensive permissions. They will act as arbitrators and decide which modifications will be applied and which will not, resulting in a secure, reliable NFR that is representative of the reality of the company's environmental sustainability. In addition, the approval of these corrections and their subsequent modification on the blockchain will allow the AI to learn and update its externality measurement systems, ensuring that the robots are reprogrammed to make better decisions and execute the necessary actions autonomously.

It is necessary to consider the use of this type of technology due to the lack of security in the information that is produced today at the time of the elaboration of NFR and the difficulties that it generates in terms of the effectiveness of its auditability. blockchains, due to their registration system, have a high guarantee of security. It is a technology recognized as immutable that guarantees the integrity of information without the need for intermediaries. In the private blockchain, the system of nodes that we propose will guarantee transparency so that the combination of the appropriate permissions together with the guarantees that the system offers, will allow a simple verification of who and when the registrations were made, significantly reducing the risks of fraud, since all members of the network have access to all the information.

The use of blockchain in the management, administration and auditing processes of NFR will revolutionize and simplify processes. The complexity of the blockchain will depend on the number of users and the load power of the network. The use of a private blockchain and a single database allows the automation of validation and verification processes to be accelerated and result in time savings and cost reductions. As for the energy problems raised by public blockchains in terms of energy consumption, private blockchains consume much less energy, since, as we have already mentioned, they require less processing capacity.

On the other hand, this type of private blockchain can be immediately deployed by an entity or can choose to access it through enterprise storage platform providers. For example, *Hiperledger* is a Linux-backed project that seeks to bring blockchain to the enterprise and industrial world, i.e., they aim to develop a common and universal platform for private blockchains.

4 Correlation of These Improvements with Environmental Taxation

If we manage to standardize and successfully implement this proposal, we move on to the next objective: to adjust tax treatment or public spending in accordance with the proposed measurement mechanisms, so that tax law is fairer and respects constitutional principles.

Focusing on current Spanish regulations (similar in many countries), the environmental taxation presents important problems in terms of its legal basis. Its classification as an extra-fiscal tax to avoid the correct justification of economic capacity is unsustainable in the long term, a. nd the risks of competition when it comes to creating, implementing and collecting this type of tax do not improve its legal framework [4].

Environmental taxation has been described mainly as extra-fiscal since its inception. This option has been chosen because penalties are considered to be too harmful. It is true that conduct that undermines sustainability and the environment is not desirable, but today it is, for the most part, tolerated by the legal system because of the benefits that certain types of activities bring to society, however harmful they may be. Since no other less burdensome measure has been found to achieve the purpose in a similar way, the creation of this type of tax has been chosen as an instrument of economic policy thanks to the usefulness of financial law as a promotional law.

The principle of economic capacity, in Spain, can be defined as the available wealth of a subject and, its importance is due to its relationship with the social solidarity that underlies the tax system, since it constitutes the criterion for the distribution of the tax burden, harmonizing economic inequality and allowing tax equality. Given its relevance, economic capacity has come to be considered: guiding principle of tax justice; material criterion for the fair distribution of taxes; criterion of unity of the concept of taxation and basis, justification and measurement of taxes.

The point of controversy in environmental taxes is the difficulty of adjusting them to the principle of economic capacity. The doctrine has defined very varied positions: some authors have understood that the deterioration of nature, as a consequence of industrial productive activities, is in itself an index of economic capacity; others believe that the "polluter pays" principle can be subsumed in the principle of solidarity as a guiding criterion of tax justice; economic capacity has also been linked to the enjoyment of the environment; among many other theories.

The correct control of the NFR, taking advantage of interactive robotics, will be able to provide reliable data. Thus, environmental taxation will achieve a higher degree of certainty on the taxable event and will allow the correct quantification of the elements subject to taxation. It will constitute a turning point that will guarantee taxation on a solid basis of data that can really bring us closer to determining a fairer and real economic capacity of the obligated entities. Moreover, it will inevitably lead us to the creation of a tax that would be in harmony with tax principles, simplify its calculation and guarantee tax justice [5].

5 Conclusion

It is necessary to consider the use of technology due to the lack of security in the information produced today at the time of the elaboration of NFR and the difficulties it generates in terms of the effectiveness of its auditability. blockchains, due to their registration system, have a high guarantee of security. This technology is recognized as immutable and guarantees the integrity of information without the need for intermediaries. In the private blockchain, the system of nodes connected to robots with artificial intelligence and the Internet of things will ensure transparency. The combination of the appropriate permissions together with the guarantees that the system offers, will significantly reduce the risks of fraud.

From this point on, it seems feasible to adjust the environmental taxation to effectively ensure the reduction of negative externalities.

References

1. DIRECTIVE 2014/95/EU OF THE EUROPEAN PARLIAMENT AND OF THE COUNCIL of 22 October 2014 amending Directive 2013/34/EU as regards disclosure of non-financial and diversity information by certain large undertakings and groups. Official Journal of the European Union. L 330/1 de 15 November 2014, pp. 1–9 (2014)
2. DIRECTIVE 2013/34/EU OF THE EUROPEAN PARLIAMENT AND OF THE COUNCIL of 26 June 2013 on the annual financial statements, consolidated financial statements and related reports of certain types of undertakings, amending Directive 2006/43/EC of the European Parliament and of the Council and repealing Council Directives 78/660/EEC and 83/349/EEC. Official Journal of the European Union. L 182/19 de 29 June 2013, pp. 19–76 (2013)
3. Proposal for a DIRECTIVE OF THE EUROPEAN PARLIAMENT AND OF THE COUNCIL amending Directive 2013/34/EU, Directive 2004/109/EC, Directive 2006/43/EC and Regulation (EU) No 537/2014, as regards corporate sustainability reporting. Brussels, 21 April 2021. COM/2021/189 final
4. García Novoa, C., Vivel Búa, M.M., Torres Carlos, M.R., Lado Sestayo, R., Aneiros Pereira, J.: Digitalización, Inteligencia Artificial Y Economía Circular, 1ª, Estudios. Thomson Reuters-Aranzadi, Cizur Menor (2021)
5. Grau Ruiz, M.A.: ¿Medidas fiscales para que la robótica contribuya al desarrollo sostenible? In: Inteligencia artificial y administración tributaria: eficiencia administrativa y defensa de los derechos de los contribuyentes. Fernando Serrano Antón (aut.) (2021)

Governance and Regulation

How to Better Involve Organised Civil Society in Regulating Robotics

C. Trias(✉)

EESC, Brussels, Belgium
Carlos.TriasPinto@eesc.europa.eu

Abstract. Beyond its main function (opinions on the EU legislation), the European Economic and Social Committee (EESC) can play a key role in involving organised civil society in the appropriate discussions and decision-making process on robotics. The EESC can enhance different ways of stimulating the participation of actors on the ground for European Union (EU) reflection and action.

1 Introduction

Complementing human capabilities through the use of **smart** and **autonomous** machines trained to allow for adopting decisions independently must be properly managed in order to avoid the risk of direct and indirect costs affecting society-related concerns.

One of the difficulties in AI & robotics is how to ensure that differing and contradictory positions are given a balanced hearing. There is no effective counterweight to the legitimate representation of the robotic sector's interests.

2 Discussions

There is a need to reach a common understanding and further harmonised definitions using a risk-based framework for categorising robotics as well as accurate procedures for assuring a right governance and effective law enforcement of those **AI applications** that should be considered **high-risk**.

Given the complexity and fragmentation of the current EU legislation, that is currently impeding a **well-functioning internal market** for AI systems, it is crucially important for civil society organisations not just to set basic objectives and call for a general tightening-up of the upcoming regulation,[1] but to put forward knowledgeable and practical proposals and arguments.

All this, in order to address concrete risk areas regarding to the scope, definition and clarity of the prohibited AI practices, the implications of the categorisation choices made in relation to the "risk pyramid", the risk-mitigating effect of the requirements for

[1] Communication from the Commission "Fostering a European approach to Artificial Intelligence". Brussels, 21.4.2021, COM (2021) 205 final. Proposal for a regulation "Laying down harmonised rules on artificial intelligence (Artificial Intelligence Act) and amending certain Union legislative acts". Brussels, 21.4.2021, COM (2021) 206 final.

M. A. Grau Ruiz (Ed.): INBOTS 2021, BIOSYSROB 30, pp. 195–198, 2022.
https://doi.org/10.1007/978-3-031-04305-5_32

high-risk AI, the enforceability of the Artificial Intelligence Act (AIA) and the relation to existing regulation and other recent regulatory proposals; as stating the reçent approved EESC opinion on the new AI regulation2.

The increasing use of AI also creates **high risks to the safety and fundamental rights** such as the opacity of many sophistication of algorithms affects security and transparency, exacerbating social inequalities and raising many ethical concerns that need to be addressed [2].

In this sense, AI-enabled robots and intelligent systems must be engineered and designed to meet the same high standards of safety and protection of fundamental rights for traditional technologies provided by European law [1].

Moreover, it would be highly recommended making third party conformity assessments obligatory for all high-risk AI.

Involving civil society organisations in testing sites and regulatory "sandboxes" will enable experimentation and piloting of new ideas whilst assuring a right governance.

Monitoring of progress should be enhanced by adding performance of Environmental, Social and Governance (ESG) indicators devised by all stakeholders.

This will require considerable efforts to ensure that the organisations can participate in discussions on **equal footing** with the legislator and with other organisations.

In this regard, the EESC is well-positioned to play a prominent role in rectifying the lack of effective civil society involvement, because the EESC includes representatives both of industry associations and of consumer organisations and trade unions.

As a European institution, the EESC is legitimized for modulating a sound **structured dialogue** between all concerned actors, respecting a balance and due representation.

Should be the starting to **civil liability**[2] issues linked to development of robotics and AI, taking advantage of remaining legislation to adapt the EU liability framework, such as revising the Product Liability Directive, in order to address liability issues related to new technologies, including AI, and by a revision of the General Product Safety Directive.

[2] Safety, health, and security; freedom, privacy, integrity, and dignity; self-determination and non-discrimination; personal data protection, etc.

It's role is to ensure privacy, diversity, non-discrimination, due process, transparency, understandability in decision-making processes, environmental and social wellbeing in alignment to the Ethics guidelines for trustworthy AI ("EGTAI").

The EU precautionary principle strongly suggest for certain decisions to remain the prerogative of humans, particularly in domains where these decisions have a moral component and legal implications or a societal impact such as in the judiciary, law enforcement, social services, healthcare, housing, financial services, labour relations and education.

3 Towards a Strengthened Framework for Co-operation

This initiative would complement the sound inputs coming from the High-Level Expert Group on Artificial Intelligence (AI HLEG) and the AI Alliance,[3] as well as the usual European Commission´s private and public stakeholder consultations and impact assessments (before launching a new rule).

As a result, a positive sum game which could considerably improve the interaction of agents to collaborative address the concrete challenges and concerns raised by the increasing use of AI.

In summary, the so called momentum for Europe taking the global lead in human-centric and Trustworthy AI, in coherence with its values, social wellbeing and environmental aspects, should go hand in hand with the establishment of an open, collaborative, and inclusive hub for AI technologies.

Could be led by the EESC and composed with partners from industry, the research and academic world, and all kind of actors such as user association representatives from civil society, from their capacity and ground experience to debating into education by incorporating ethical and environmental issues, lifelong learning as well as skills development and support for workers and businesses.

Further implementation of new concerned **rules** and **codes of conduct** (self-regulation) could also follow the same scheme in terms of dialogue and joint commitments.

In turn, the **multi-level governance** approach guarantees that power is spread **vertically** between many levels of government[4] and **horizontally** across multiple quasi-government and non-state actors that are mobilized in policymaking. In our view, this model constitutes the adequate tool for ensuring the right participation of all concerned actors, depending on related issues.

Those stakeholders involved in the **co-creation**, development and commercialisation of robotics should guarantee their **business models** are built in security and ethics of the technology they produce. They must also be monitored following appropriate regulatory standards.

As a key successful factor**, data access and privacy concerns** need to be addressed, assuring **accountability** with databases without human intervention. Verification needs

[3] The AI Alliance is a multi-stakeholder forum launched in June 2018, AI Alliance https://ec.europa.eu/digital-single-market/en/european-ai-alliance.

[4] Crosses the in practice separate domains of EU member States and their regions and EU institutions.

to be built into the process of automated and algorithmic decision-making, showing **full transparency** in order to guarantee a **human-centred approach** that contemplates the **European values** regarding fundamental rights as well as the **SDGs** in their purposes.

Blockchain[5] demonstrates that by combining peer-to-peer networks with cryptographic algorithms a group of actors can reach an agreement on a particular state of affairs and record that agreement without the need for a controlling authority.

The combination of blockchain with other distributed systems, such as **robotic swarm systems**, can provide the necessary capabilities to make robotic swarm operations more secure, autonomous, flexible and even profitable.

References

1. Nagenborg, M., Capurro, R., Weber, J., et al.: Ethical regulations on robotics in Europe. AI & Soc. **22** (2008)
2. Stutis, M., Gordon, J.-S.: Smart Technologies and Fundamental Rights. Brill, Leiden (2020)

[5] Emerging technology, originated in the Bitcoin field, that allow to trace a transaction.

Can the Legal Requirements Foreseen for Other Sectors Aimed at Protecting the Market and, Ultimately, the People and Users Themselves, Be Transferred to the Robotics Industry Sector?

Natalia Zardoya Jiménez[1](✉) and Christi Amesti Mendizábal[2]

[1] Law and Business Administration, Complutense University of Madrid, Madrid, Spain
nataliazardoyaj@hotmail.com
[2] Commercial Law, Complutense University of Madrid, Madrid, Spain
camestim@ucm.es

Abstract. Artificial Intelligence (AI) and robotics are two realities whose impact and benefits on the economy and society are growing. However, we are dealing with intelligent machine with behaviour that can be harmful and, in some cases, unpredictable, which implies a great risk that could be considered as unacceptable in specific situations. This is precisely why it is necessary to analyse the legal and ethical consequences of the use of intelligent robots in our society, as well as establishing possible solutions based on existing legislation in other high impact areas.

1 Introduction

Robotics, and all that this new field entails, poses new challenges in terms of safety and guarantees to be provided to users of these products. Indeed, from the point of view of business regulation, robots, and artificial intelligence in general, will require a legislative effort and a correct vision for an adequate application of the law.

In this regard, it should be noted that it will have to be studied whether the current legislation applicable in the Member States is sufficient and provides coverage to deal with the risks and new situations and events that may arise when the use of this type of "products" spreads in the market and are used by natural or legal, public or private persons.

On the other hand, the aspect of the administration of the companies and management of the companies dedicated to this type of activities or that use this type of products in their production or organization must be emphasized, becoming an important issue in several areas, especially in matters of liability of the employer, both in the civil and administrative area and even in the criminal area. Likewise, the importance in this area of Corporate Social Responsibility should be highlighted as offers a model of administering and managing the new situations that arise, as well as the drafting and compliance with the ethical codes that formalize these companies in this sense.

Jurists should study whether the Spanish legal system and the European legal system in general, are prepared to solve the commercial-business problems that may be generated

M. A. Grau Ruiz (Ed.): INBOTS 2021, BIOSYSROB 30, pp. 199–206, 2022.
https://doi.org/10.1007/978-3-031-04305-5_33

by robots, especially those of physical-cognitive interaction with humans. In this sense, and among other aspects, the problems of liability for damages caused to users and other persons close to the robot, as well as to other goods, should be studied. For instance, would the manufacturer or importer of the robot be liable. This is currently addressed in Spanish law by the Royal Legislative Decree 1/2007, of November 16 (LGDU), which approves the Revised Text of the General Law for the Defense of Consumers and Users and other complementary laws, with respect to Defective Products and in the Civil Code with respect to contractual and non- contractual liability.

In this respect, it may be of interest to make a brief and succinct review of the special measures contemplated in other business sectors, in order to grant greater protection to the consumer of their products or services. In this regard, and in particular in this article, we will refer to the banking sector.

On the other hand, robotics and artificial intelligence is a sector that is taking a leading role of great relevance in the management of any business sector, including banking itself, and other areas of public or private nature, such as public administration, health, state security, culture, education, areas in which today it is essential to incorporate data with different intentions and of different character, a very sensitive issue for the privacy of individuals and for decision making of any nature.

Therefore, it is necessary that there is a control in all those aspects of the activity that may involve the introduction of data, being of singular importance that there is a system that supervises not only the veracity of the data, but the relevance, possible risks, fairness and non-discrimination of human, social and political rights in the incorporation of such data in robots.

2 The Correct Supervision of Robots

As mentioned above, one of the most regulated industries due to its great impact on our society is the banking industry. In fact, it should be noted that banks are an essential element of a country's economy because they manage citizens' savings and finance entrepreneurs and individuals, as well as the States themselves, being essential for the functioning of a country.

Although the robotics sector and the banking sector are very different, certain similarities can be discerned in the way in which measures are articulated to monitor the different risks they generate.

The first, the unique system of banking supervision should be highlighted. In the case of the European Union, since 2014 the supervision of credit institutions has been the responsibility of the single supervisory mechanism, a system led by the European central bank and in which the national supervisory authorities participate, such as, in the case of Spain, the Bank of Spain [3]. The reason for the existence of this complex framework is the evolution of the banking system, with the proliferation of complex financial instruments and the diversity of factors influencing the recurrent activity of institutions. This gives rise to the existence of multiple risks, the management of which by institutions must be properly regulated and supervised, in order to guarantee the stability and security of the system [3].

In this sense, robots equipped with artificial intelligence, depending on the uses or purposes for which they are intended, may entail a series of risks, with repercussions such as the violation of fundamental rights and/or safety standards [10]. As highlighted in the White Paper on Artificial Intelligence, "these risks may stem from flaws in the design of AI technology, be related to data availability or data quality issues, or others arising from machine learning". The importance of risks in AI is such that the supervisory framework proposed to be designed by the European authorities is based precisely on a risk-based approach [4]. This approach has been adopted by the European Parliament and the Commission in their respective proposals for regulations, namely in in the European Parliament Resolution of 20 October 2020 on Recommendations to the European Commission containing a proposal for a regulation on a framework for ethical aspects of AI, robotics and related technologies and the subsequent proposal for a Regulation containing the so-called Artificial Intelligence Act of the European Commission published on 24 April 2021. In the latter, the Commission proposes a regulatory framework in proportion to the risk of AI in each of its applications, thus restricting the most restrictive rules to AI considered to pose the most significant risk, paying particular attention to consumer protection and fundamental rights. "*To that end, the draft AI act distinguishes between AI systems posing i) unacceptable risk, ii) high risk, iii) limited risk, and (iv)low or minimal risk*" [5].

An example of a high-risk use would be the use of AI in recruitment processes. If, for example, an intelligent robot were to sift through a company's CVs or even interview candidates for a specific job, depending on the data that has been previously entered, as well as the characteristics of the business environment in which it operates, there may be a greater or lesser risk of maintaining or even increasing harmful situations such as the "gender gap" [3].

Therefore, for this and other situations, appropriate supervisory and regulatory mechanisms will have to be designed. In this respect, the Artificial Intelligence Act proposes takes on board the ideas of the White Paper on Artificial Intelligence and regulate a European governance structure where authorities acting at EU level, as well as at national level, converge. It would also include mechanisms for the participation of civil society and stakeholders. In this respect, national supervisory authorities would act as first points of contact in case of suspected non-compliance with the legal obligations and ethical principles included in the Resolution itself. Ethical principles refer to: "*safety, transparency, accountability, non-bias or non-discrimination, social responsibility and gender equality, right to redress, environmental sustainability, privacy and good governance*". The national supervisory authority must carry out "*an impartial, objective and external risk assessment*" in accordance with the criteria set out in the proposed Regulation. This governance structure would apply not only to robotics, but also to artificial intelligence and other related technologies [5, 7].

These draft leaves several noteworthy ideas on how to approach robot governance. The first is the need for coordinated action, given the cross-cutting nature of IA and the implications for the single market, as is the case for the supervisory mechanisms of the banking sector. In this respect, despite the added value this entails, the distribution of competences between national and European authorities should be comprehensively determined, in order to have a coherent and orderly mechanism. At the same time, the proposal highlights the importance of the exchange of views with stakeholders and civil society is properly underlined. This would allow a closer look at the problems and proposals within the AI and robotics sectors, avoiding that unsatisfactory and inadequate measures could be implemented. Finally, it also highlights the need to assess and monitor the risk of non-compliance not only with laws but also with ethical principles, and further specifies what these principles would be. On the latter, principles such as fairness, precisely to mitigate risks associated with the situation described above, stand out in the use of AI in procurement processes [5]. These and other principles can give new meaning to the application of existing texts on AI-focused topics, thus contributing to a more faithful interpretation of our new environment, and can be a source of inspiration, a necessary regulation, which some are already globally calling "Robot Law" [2].

3 Consumers Protection

Another of the characteristics of this supervision system is the search for human participation, so as to ensure that the different situations in which AI is involved are provided with human control and oversight, to a greater or lesser extent, depending on the risk they may entail [4].

However, this idea is not easy to put into practice, since it does not seem that the entities that manufacture and use robots can be limited. In fact, it must be said that establishing constraints to the entry of new companies and, probably, of new inventors, would be in addition to "putting doors to the field", limiting the advance and progress of technology and science, which can hardly be justified, and even cannot, nor should be done, except in clear cases [3].

The question should therefore be asked as to how a system of supervision by individuals could be set up, which would be sufficiently safe but at the same time respect and encourage the innovation inherent in this technology. As mentioned above, consumer protection would be at the heart of the regulation of AI and robotics, just as it is also a fundamental pillar of the supervision of the banking sector. Precisely, the example of the banking sector is useful to understand how consumers and users are subjects that are afforded special protection due to their situation of vulnerability, compared to other actors with greater economic power or protected by other legal forms. Furthermore, it is worth noting the special measures that are envisaged in order to provide greater consumer protection with regard to the products or services offered by financial institutions. Indeed, in the banking sector, as indicated above, in addition to demanding legislation regarding compliance by banking entities with special rules on accounting and security and economic-financial stability, they are under the supervision of the corresponding National Banks (Bank of Spain) and the European Central Bank [3].

In the case of intelligent robots, the consumer market is key to the development of robotics, with a multitude of intelligent and non- intelligent robots being sold as consumer goods, making consumer data the most compromised and exposed to cyber-attacks [9].

The question arises as to whether the existing and proposed mechanisms would be sufficiently respectful of the principles of progress and innovation, while at the same time guaranteeing the safety and protection of consumers and users in contact with intelligent robots. For this purpose, it is necessary to know the state of consumer protection regulation for the specific case of intelligent robots.

Spanish and European regulation on this topic include liability for damages in their articles, in order to protect them from possible damages associated with the products they obtain from manufacturers. However, two questions arise around this, precisely because of the uniqueness of the potential risks of intelligent robots.

Firstly, as mentioned above, robots may manifest behaviours in unexpected ways. Secondly, the figures and actors involved in the operation of the robot are very varied: from the designers and manufacturers to the users themselves who interact with the robot. The intervention of different parties in the aforementioned areas makes it difficult to apply the rules governing non-contractual liability, which is understood as that which responds to the damage produced independently of any pre-existing legal relationship between the parties [3].

The Spanish law regulating this matter (the LGDU) restricts liability for damage caused by defective products to the producer (art.135, LGDCU). Furthermore, the same regulation defines a defective product as [one that does not offer the safety that could legitimately be expected (...)] (art.137, LGDCU). This liability means that in the event of fault or negligence on the part of the producer, the latter must be liable for the damage caused, provided that the direct causal link between the defect and the damage is proven, and this task corresponds to the injured party in question (art. 139 LGDCU).

When analysing this regulation in the light of the singularities of the operation of the robots mentioned above, several problems arise. For example, it should be considered whether, from a legal point of view, a robot that acts autonomously, outside the functions predetermined by the manufacturers or developers, could be categorised as defective. One of the main questions that arises from a legal point of view is, in our opinion, whether it would be possible to categorise as defective a robot that has carried out an act of an injurious nature, in some unforeseeable cases, and which is based on autonomous learning that is outside of what, in the first instance, was established by the manufacturer or developer of the programme itself [3, 6].

This problem responds to the lack of acceptance by the regulations of the new technological reality that is breaking with force within the different productive sectors, so we need new instruments that can embrace these changes and that serve as a complement to the existing ones. This is even more important when it comes to AI monitoring mechanisms such as the one described above, when talking about the Artificial Intelligence Act. The Commission proposes that national consumer protection authorities should be involved in the process of assessing the proper development of AI, and it is therefore absolutely necessary for them to be equipped with the appropriate legal tools to enable them to resolve disputes in an appropriate manner. [5, 7]. In the same way, legal uncertainty may reduce overall levels of security and undermine the competitiveness of European companies [4].

In an attempt to find a solution to the previous problem, controversial solutions have emerged, such as the one proposed by the European Parliament in 2017, which consisted of creating a specific legal status for robots by creating a electronic personality [4]. However, this idea seems to be far removed from the recent positions of both the Parliament and the Commission, as described above when analysing their respective documents on a supervision mechanism. I would also add that the background to this idea is also fraught with questions and difficulties, not only legal, but above all, in our opinion, ethical and moral.

Indeed, the European regulator seems to be more inclined towards control over AI systems, and in particular human control that considers ethical principles and the greater or lesser likelihood that these may or may not be respected [8].

Although the latest proposals for a Regulation do not mention changes to consumer protection legislation, they do describe new lines of action aimed at increasing confidence and safety in AI products. On the latter, it is necessary to elaborate further and to know what the concrete measures are according to each level of risk. Among the most important measures is the *"requirement of an ex-ante conformity assestment"* for the high-risk AI systems. Through this measure, an EU-wide database would be created, and providers of high-risk AI systems shall be required to register their systems on it. Also, for this type of technologies, it will be required to comply with several conditions relating to:

"Testing, technical robustness, data training and data governance, transparency, human oversight, and cybersecurity" (Articles 8 to 15 of the Artificial Intelligence Act) [5].

In this way, the new regulatory framework aims to create safe and trustworthy AI systems and defining mandatory requirements applicable to the design and development of AI systems before they are placed on the market [1].

4 Conclusion

It is clear, from the direction that recent regulatory proposals are taking, that AI is increasingly comparable in terms of scope to sectors such as banking. We will have to await the approval of the final regulation by European regulators, but the creation of a mechanism at European level would undoubtedly be a good starting point, given the advantages that this would bring for the proper functioning of the digital single market and the correct exploitation of these technologies. It could also enhance the EU's leadership in the development of AI, as has already happened regarding the protection of personal data. Also, maintaining and nurturing the anthropocentric stance of the human being throughout the deployment of AI, together with the creation of common ethical principles, is not only not an impediment to AI, but also contributes to its proper development.

However, despite the novel and, a priori, positive proposals, it should not be forgotten that it runs the risk of being a hindrance to innovation and deployment. Such a system in which there are tools that limit to a certain extent its use, that impose administrative formalities to be able to commercialise certain technologies and with a fine policy attached, runs the risk, if it is not implemented with sufficient efficiency and agility, of putting up walls to development, thus acting to the detriment of European companies compared to other international leaders [8]. However, the regulation identifies most existing AI as limited, minimal and zero risk, and therefore would not apply limitations of the same importance as those that may pose a higher risk [5]. On the other hand, the existence of a mechanism such as the one established in the draft regulation of the Commission, presents the risks inherent in any such large-scale scheme, but given the example of other existing large-scale schemes such as banking, a design with the least possible dysfunctionalities may be more feasible.

In this regard, the Commission has already announced that the Directive on liability for damages will be revised to include regulatory developments concerning AI and other technologies [1]. The reason for this is that even though the report analyses major issues surrounding AI and related technologies, leaves unresolved questions such as the ones we have raised, as well as imposing measures which, for some European bodies, are still not sufficiently guaranteeing. In fact, European Consumer Organisation, BEUC, argues *"that the proposal should have a broader scope and impose basic principles and obligations (e.g. on fairness, accountability and transparency upon all AI systems, as well as prohibiting more comprehensively harmful practices"* [11]. As a result, we will have to keep an eye on developments in this area.

In conclusion, we see how the measures in place for other sectors such as banking have certain similarities and their example can help to shape a more robust system for AI. Likewise, although with necessary adaptations, consumer law is a direct protection tool for the consumer in the event of defects in an intelligent robot. While it is true that, as has been pointed out, existing measures in this area need to be improved, the Commission's new proposal already establishes a control mechanism prior to any defect. In other words, the regulator has chosen to adopt measures in the regulation of the product creation, development and deployment phase, as opposed to de facto amending existing consumer protection legislation.

Finally, we hope that the machines we have created in our image and likeness, even if they displace certain human abilities, will never displace the particularities of the human person, but that we will always set the rules and limits while respecting our own uniqueness, making robots an instrument of help and of which we can avail ourselves. Finally, in no case should it destroy or displace the best of our way of life, in short, the best of ourselves.

Acknowledgment. This project has been carried out within the framework of the European project INBOTS, which has received funding from the European Union's Horizon 2020 research and innovation programme under grant agreement No. 780073.

References

1. Anguita Ríos, R.M.: Inteligencia artificial y derecho civil: líneas de pensamiento en materia de daños. Rev. Crítica Derecho Inmobiliario **781**, 2541–2581 (2020)
2. Aransay Alejandre, A.M., et al.: Derecho de los robots, 1st edn. Wolters Kluwer, Madrid (2018)
3. Mendizábal, C.A., Jiménez, N.Z.: El buen gobierno de los robots para su correcto impacto en la sociedad: ¿qué herramientas existen?: Rev. Arbor **802** (2021)
4. European Commission, White Paper on Artificial Intelligence - A European approach to excellence and trust, Brussels: COM (2020). 65 final
5. European Commission Proposal for an Artificial Intelligence Act, Brussels: COM (2021). 206 final
6. Lacruz Mantecón, M.I.: Inteligencia Artificial y coches autónomos. Análisis jurídicos europeos. Rev. Crítica Derecho Inmobiliario **775**, 2373–2409 (2019)
7. Grau Ruiz, M.A.: Nuevas tecnologías disruptivas y tributación, chap. 2, p. 262. Aranzadi, Pamplona (2022)
8. de Miguel Asensio, P.A.: Propuesta de Reglamento sobre Inteligencia Artificial. La ley Europea **92**, 1–8 (2021)

9. European Parliament. A comprehensive European industrial policy on artificial intelligence and robotics, Strasbourg (2019)
10. European Parliament. European Parliament resolution of 20 October 2020 with recommendations to the Commission on a framework of ethical aspects of artificial intelligence, robotics and related technologies (2020)
11. European Parliamentary Research Service, Artificial intelligence act: EU Legislation in Progress (2021)

Perspectives from the Legal Practice: Robotics

Santiago Mediano(✉)

Presidente de Santiago Mediano Abogados y Presidente de la Sección de Robótica, IA y Realidad Virtual y Aumentada del Ilustre Colegio de Abogados de Madrid (ICAM), Madrid, Spain
smediano@santiagomediano.com

Abstract. The main debate includes perspectives from the legal practice and robotics. How could we achieve alignment between technological development and respect for human rights?

1 Introduction

The arrival of robotics and artificial intelligence in the legal world is now a fact and presents new (and important) challenges that, among other things, have led to the emergence of a new branch of law: Robot Law.

Robot Law may seem to be a futuristic or unrealistic concept, but just as the Internet was at its time, robots and artificial intelligence (AI) are the game-changing technology of our time. Robots are part of our daily lives, and they are bringing about substantial changes to our society: we will see robots performing all kinds of tasks, from driving trucks to improving the work of many professionals —healthcare, physically demanding jobs— or even replacing and collaborating with humans in situations of physical difficulty or in natural disasters. They are also replacing conventional household appliances to create interconnected, smart homes that make life easier for humans.

It is therefore essential that the law deals with the regulation of this disruptive technology to ensure respect for the fundamental rights of the constitutions of all states worldwide.

How could we achieve alignment between technological development and respect for human rights?

Alongside the great enthusiasm for the promises and established personalities of the electronic field, a dual legal concern arises. On the one hand, the erosion of fundamental values such as dignity, autonomy, privacy and non-discrimination, which could be undermined by such developments. And on the other hand, the threat of technology becoming another way of making people less human and more passive, with reduced autonomy. In other words, the imperfect essence of humanity could be lost through technological development.

2 Perspectives from Legal Practice

2.1 Sophia the Robot and the Metaverse

Sophia the robot is a "humanoid" created in 2016 that has been considered the world's most advanced artificial intelligence (AI). In March 2021 auctioned its own non-fungible

M. A. Grau Ruiz (Ed.): INBOTS 2021, BIOSYSROB 30, pp. 207–211, 2022.
https://doi.org/10.1007/978-3-031-04305-5_34

tokens (or NFTs) in conjunction with Italian artist Andrea Bonaceto, at the height of this trend.

In 2022, Sophia is joining the metaverse. A virtual anime version of Sophia will be tokenised and auctioned as part of a promising new ecosystem project nicknamed "Noah's Ark".

The collection will be called "The Transmedia Universe of Sophia beingAI" and, as part of the partnership, all NFTs will be supported by Alethea AI's decentralised Noah's Ark Metaverse project.

2.2 Tesla Bot

Tesla Bot is the new humanoid robot created by Tesla and is designed to replace humans in certain tasks, specifically, and according to its creators, "the most risky, repetitive and boring ones". So far, Elon Musk has presented only a mannequin that demonstrates that the project is much more than a series of formulas on paper. It looks very similar to a human: it is 1.76 m tall and weighs 56 kg. It is said to be able to move loads of up to 20 kilos and can travel at a maximum speed of 8 km/h.

2.3 Transhumanism: Politics of Hybridizing Humans with Robots

For several centuries, political movements have emanated from nationalist or religious convictions, socio-political and economic theories, or even environmental concerns. However, scientific and technological issues have rarely driven the dominant socio-political discourse, as scientists have hardly engaged in proposing social constructs for the future. However, a technology-oriented intellectual movement could soon become the focus of political debate, as 21st century science is on the verge of reshaping human identity itself.

By the turn of the millennium, transhumanism emerged as a futurist social movement, led by some of the world's leading researchers and heads of global technology corporations (mostly based in the super-rich Silicon Valley), but has only recently gained wider media coverage and popularity, as well as inspiring the plots of many Hollywood movies.

In simple terms, transhumanism wants technology to be allowed to enhance human capabilities, something often denied by "conservative or ill-informed" electorates and politicians.

Claiming to represent the scientific community as a whole, transhumanists seek to pursue controversial technologies related to gene editing, artificial intelligence, etc., which are often subject to regulation as violating certain ethical or religious sensitivities.

In the words of Zoltan Istvan, the leading transhumanist who ran for the US presidency in the 2016 election, "scientists and technologists are usually not the loudest when it comes to politics, and yet if you don't approve of certain kinds of regulations, you can't practice science. I mean we have endless debate about gene-editing technology, about whether we should regulate the field of artificial intelligence, but most scientists don't step forward to voice their opinions or send lobbyists to Washington, so that's what the transhumanist party is trying to do".

As the influence and benefits of biotechnology, nanotechnology, robotics, etc. increase at an exponential rate, transhumanists advocate pouring more money and resources into the development of new technologies that promise human longevity and increased capabilities by hybridising the human body with artificially intelligent robots and the internet, thus creating a new race of superior humans called "transhumans".

Thus, the three stated goals of transhumanism are "super longevity", "superintelligence" and "super well-being", although smartphone pioneer David Wood has recently added "superdemocracy" to this list.

While there is nothing new in the use of technological substitutes for human body parts, the current debate against transhumanism centres on the question of enabling implants that augment human sensory perceptions and enhance cognitive ones by connecting human bodies directly to computers via chip implants.

2.4 Human Rights in the New Wave of Transhumanism

In Aldous Huxley's futuristic novel "Brave New World", we were introduced to "London's incubation and conditioning centre, where children are created out of the womb and cloned to increase the population". At the time, the centre was just an unattainable ideal, but in today's times, it is not as improbable.

How did reproductive science go from fantasy to prediction? In vitro fertilisation has been a reality for almost four decades, while altering the DNA of unborn babies has only recently become possible. This advance in science is accepted by advocates of transhumanism such as Julian Savulescu, who sees it as a "moral duty for future generations". Melinda Hall, a philosopher, explains her views in the following terms: "Its moral principle of procreative beneficence urges those considering reproduction to apply evolving technologies as they emerge to create the best possible child with the greatest future".

Technological advancement undoubtedly brings challenges to contemporary human rights and not exclusively in the field of life, but also impacts on death. Take, for example, the thousands of people who since 1988 have trusted companies to be "hosted" in cryonics centres after their death. What will happen when these companies finally develop a technology to "resurrect" these humans, will they be given a new identity, will they be entitled to a "clean slate" of their lives, what will happen to their past life properties, what if they committed a crime for which they have not been tried?

More questions continue to arise when we visualise the image of this resurrected human that is full of technological enhancements such as brain implants, it makes a reasonable mind wonder: will the companies that develop these technologies be able to manipulate their customers, where will there be a version without advertising, who will own the implants, who owns the data that transits the implants, who owns the data that transits the implants, who owns the data that transits the implants? These are some of the new realities that human rights must face. "The dogma of human rights was formed in previous centuries as a weapon against the Inquisition, the ancient regime, the Nazis and the Ku Klux Klan. It is hardly equipped to deal with superhumans, cyborgs and super-intelligent computers".

Questions have arisen that cover essential aspects of human rights law. These questions cover a range of issues, from the relevance of its legal infrastructure to how to respond to its changing philosophical foundations. Some critical questions to be assessed are: when does a human being cease to be human; is human consciousness sufficient to have rights; when does the application of human rights end; what are the limits, if any, of technological enhancements; does freedom have limits when it comes to enhancing human capabilities; are human rights instruments applicable to these new realities; will a fourth wave of transhuman rights emerge; and will a fourth wave of transhuman rights emerge?

But of all the questions, the most momentous is whether this potential fusion with machines will be liberation or enslavement. Time will tell the answer.

2.5 Three Paraplegics Walk Again One Day After Receiving Electronic Implants

Neuroscientist Grégoire Courtine, from the Swiss Federal Institute of Technology in Lausanne, Switzerland, and neurosurgeon Jocelyne Bloch, from the university hospital in the same city, lead the scientific team responsible for this recent achievement. They succeeded in implanting "electrodes that emit synchronised electrical pulses that mimic the signals that circulate along the spinal cord, which links the brain to the lower limbs". In turn, the electrodes are connected to a computer with an artificial intelligence system that reproduces the impulses needed to walk, ride a special bicycle or paddle a canoe (in the case of a patient with no lower thoracic mobility).

Courtine's professional team has been trying for years to restore mobility to people made paraplegic by accidents. In 2014, he tested his system on mice that had previously had their spinal cords severed, and two years later he did the same with monkeys. In autumn 2018, the Swiss team presented the innovations with David Mzee, a young man who became paraplegic at the age of 20. Thanks to this type of epidural stimulation and with the help of a walker, Mzee was able to walk again.

2.6 The Phenomenon of Self-driving Cars

Self-driving cars have been the future of mobility for years and the need for their development seems set to continue to grow, to the extent that in some cases the vehicles themselves will be real information servers on wheels.

When people talk about the cars of the future, it is normal to mention Tesla, one of the car companies that has made the greatest efforts to modernise the sector. However, autonomous driving is the goal of many other companies that are working every day to make the car intelligent. Taking of course the concept of "intelligent" to mean a car that really knows how to drive itself.

So, at this point, what do you think the cars of tomorrow need to be truly self-driving cars?

From a legal perspective, although the autonomous vehicle improves road safety and increases overall traffic efficiency, the underlying technologies in intelligent vehicles still need to evolve for their effective and widespread use on the roads. Legislation must enable innovation while ensuring safety and cybersecurity for drivers or occupants of vehicle seats. This is really yet another challenge for the institutions regulating the approval of autonomous technologies in vehicles. For the time being, legal experts are already reflecting on questions such as the existing levels of autonomy of vehicles, the applicable liability regime, the liable parties to be compensated for the damage caused.

3 Conclusion

The emergence of new digital technologies such as robotics or AI pose new challenges in terms of product safety, liability or data protection, such as connectivity, autonomy, data dependency, opacity, complexity of products and systems, software updates and the increased complexity of managing security and value chains.

As humanity stands on the threshold of an era in which robots, bots, androids and other increasingly sophisticated forms of artificial intelligence look set to trigger a new industrial revolution —likely to affect all strata of society—, it is important to join forces to regulate Robot Law to enable technological development while respecting human rights.

Check for updates

LIAISON: Liaising Robot Development and Policymaking to Reduce the Complexity in Robot Legal Compliance

H. Drukarch(✉), Carlos J. Calleja Ahmad, and E. Fosch-Villaronga

eLaw Center for Law and Digital Technologies, Leiden University, Leiden, The Netherlands
{h.g.drukarch,c.j.calleja.ahmad,
e.fosch.villaronga}@law.leidenuniv.nl

Abstract. The relationship between robots and policy development is complex. Technology and regulation evolve, but not always simultaneously or in the same direction. At the same time, robot developers struggle to find suitable safeguards in existing norms applicable to them. This often results in disconnections between both worlds. New robots and applications may not fit into existing (robot) categories (a robotic garbage collector or a robotic wheelchair with a robotic arm with a feeding function). Also, regulations may be hard to follow for developers who are concerned about their particular case because legislation (private and public policy making) may be outdated and with confusing types (such as 'personal care robots' not for medical purposes from ISO 13482:2014), and technology-neutral. Since legal responsiveness does not always follow as a consequent step in response to technology's dramatic pace, we initiated the LIAISON Research Project. LIAISON follows the ideal that lawmaking 'needs to become more proactive, dynamic, and responsive' to achieve its desired policy goals and explores to what extent compliance tools could be used as data generators for robot policy purposes to reduce the complexity in emerging robot governance, and unravel an optimal regulatory framing for existing and emerging robot technologies.

1 Introduction

Despite their tremendous and until recently unimaginable potential, new technologies sometimes question and challenge existing norms, breathing into existence the need for legal change. While technology's pace dramatically accelerates; however, understanding technology's implications and consequent assessment and legal responsiveness do not always follow in parallel [1–3]. Consequently, an increasing gap has come to exist between the policy cycle's speed and technological and social change [4], particularly in the field of Artificial Intelligence (AI) and robotics, i.e., movable machines that perform tasks either automatically or with a degree of autonomy.

This contribution was written as part of the PROPELLING project, an FSTP from the H2020 Eurobench project, which has received funding from the European Union's Horizon 2020 Research and innovation programme under grant agreement No 779963.

M. A. Grau Ruiz (Ed.): INBOTS 2021, BIOSYSROB 30, pp. 212–219, 2022.
https://doi.org/10.1007/978-3-031-04305-5_37

Moreover, because legal responsiveness does not always follow technological developments as a consequent step [1–2], current legislation also lags and does not frame technology adequately [5]. Consequently, current developments fail to ensure an adequate level of safety and do not perform well in the wild [6]. Given that robotics is a realm characterized by uncertainty on the risks and appropriate safeguards for covering them, there is an urge to establish a coordinated regulatory front matching interests that can easily translate into concrete, practical, and widely adopted actions for making robots safe to use [7].

Compliance tools may offer a valuable tool to help developers in their efforts towards robot legal compliance. However, new applications may fail to fit into existing (robot) categories, and regulatory frameworks might be outdated, contain confusing categories, or directly be technology-neutral, which may be hard to follow for robot developers concerned about their particular development [5]. Bearing this in mind, the LIAISON Project, a subproject from the H2020 COVR Project, investigated to what extent compliance tools, such as the COVR Toolkit, could serve as data generators for emerging robot governance purposes. To this end, LIAISON focused on personal care robots (ISO 13482:2014), rehabilitation robots (IEC 80,601–2–78–2019), and agricultural robots (ISO 18497:2018).

This chapter brings into view the inner workings and some of the critical findings of the LIAISON project. Following a brief introduction, Section II frames the inefficiency and inadequacy of emerging robot governance. Section III introduces the LIAISON model and methodology, followed by an overview of the essential findings and lessons learned from building a dynamic framework for evidence-based robot governance as derived in Section IV's LIAISON project. Finally, the paper concludes with a summary of the achieved results and their potential impact on the governance of emerging technologies.

2 A Legal 'Tragedy of the Commons'

Regulation is often very complicated, calling for a delicate interplay between various constraints, including the plurality and de-centeredness of our legal systems, the unclear fit of the regulated reality to the new development, and the unforeseeable impacts of emerging technologies. The realm of robotics is not an exception, with many developments (e.g., automated vehicles, additive manufacturing, collaborative robots/systems, or robots outside the industrial environment) still largely unregulated [8]. In this sense, a recurrent question is how the law keeps up with such technological advances.

As a result of the inability of regulators to keep up with the fast pace of technological innovation, private actors have increasingly taken the lead to develop standards that aim to mitigate the ethical and legal risks and concerns posed by robotics, implying a shift in the centralization of regulation. Nevertheless, the speed with which robotics advances and the absence of a formal communication process between robot developers and regulators from which policies could learn significantly hinders the understanding of the need to create technology-specific, long-lasting, and adequately fitting benchmarks, both for private and public policymakers. Moreover, harmonization of existing standards at a European level is significantly lacking, and even in those areas falling within existing regulations, the law often fails to guide its addressees' behavior properly. In light of all the complexities and challenges robot technology brings to the surface, several initiatives have been taken so far calling for issue managers in the form of Governance Coordinating Committees [9], Agencies for Robotics, and AI [10], or emerging technology policy labs [11]. However, a backstep mechanism that can coordinate and align robot and regulatory development in robot governance has long been overlooked, and a more thorough understanding of how policies can frame robotics and AI will be necessary to avoid potential mismatches with state of art, enhance innovation, and protect user rights [7].

3 Liaison: The Model

In 2018, Fosch-Villaronga and Heldeweg built on and complemented earlier initiatives by proposing a governance process that can serve as a backbone in coordinating and aligning robot and regulatory developers [7]. Whereas juridical analysis commonly follows top-down approaches to address legal and ethical aspects for robot technologies [12–13], this theoretical model envisions an iterative regulatory process for robot governance, which is hybrid and combines top-down and bottom-up approaches, encouraging the use of field knowledge to assist in identifying gaps and inconsistencies in frameworks governing relevant technologies [14].

Based on the conviction that lawmaking 'needs to become more proactive, dynamic, and responsive [15], LIAISON proposes the formalization of a communication process between robot developers and regulators from which policies could learn and investigates to what extent compliance tools could be used as data generators for robot policy purposes to reduce emerging robot governance complexity [16].

As such, LIAISON tests the already theoretically defined iterative regulatory process for robot governance in practice to better coordinate and align robot and regulatory development [16]. In particular, it conceives an effective way to clarify what regulatory actions policy and standard makers should take to provide compliance guidance explain unclear concepts or uncertain applicability domains to improve legal certainty. It does so by extracting compliance and technical knowledge from the COVR Toolkit, developed by the H2020 COVR Project, and directing this data to policy and standard makers to unravel an optimal regulatory framing, including decisions to change, revise, or reinterpret existing regulatory frameworks for existing and emerging robot technologies [16] (See Fig. 1). More specifically, LIAISON used desk research to investigate robots' safety compliance landscape and refined the findings resulting from this research through exploratory meetings with robot developers and relevant policymakers (including the EC, ISO and NEN, the Dutch Chapter of ISO) and feedback from the broader community of stakeholders as retrieved through a set of feedback loops (i.e., online surveys) distributed across the networks built by the Digital Innovation Hub (DIH) on Healthcare Robotics, the DIH Agrobofood, and the H2020 COVR network, and were posted on social media platforms (i.e., Twitter and LinkedIn). To this end, LIAISON focused on personal care robots (ISO 13482:2014), rehabilitation robots (IEC 80601–2–78–2019), and agricultural robots (ISO 18497:2018), thereby aiming to uncover regulatory gaps and inconsistencies and, additionally, gain insight into the usefulness and perceived value of the novel robot governance mechanism it puts into practice.

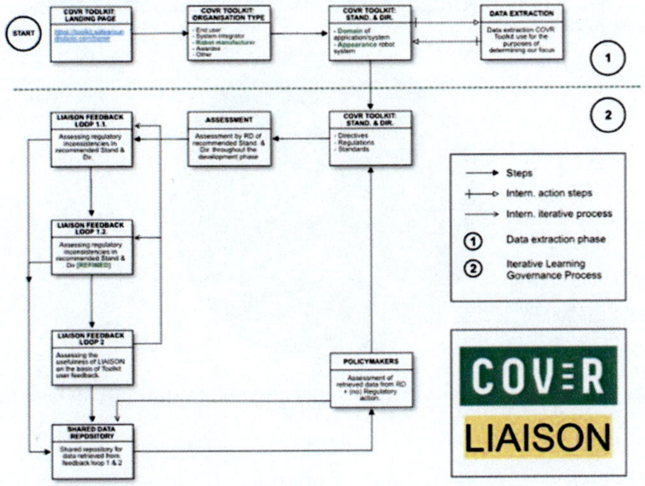

Fig. 1. LIAISON model for an iterative regulatory process for robot governance

4 Key Findings, Lessons Learned and Discussion

The LIAISON Project has, first and foremost, highlighted that engaging robot developers and end-users are essential in identifying unregulated and underestimated challenges that should be covered and addressed by robot regulatory frameworks [16]. However, although the domain of robot regulation involves a large ecosystem of public policymakers, standardization organizations, robot developers, manufacturers, and end-users, this ecosystem currently does not have a common platform for better-aligned interaction [17]. Over time, failure to create such a platform may create power imbalances concerning the creation and production of norms geared towards framing robot development and user rights protection. LIAISON has highlighted how reshaping the mode of interactions yields helpful feedback from the community and thus facilitates the iterative regulatory process [17], bringing all stakeholders together to align their interests into making current and future robot technologies safe to use will be essential.

The results gained throughout the LIAISON Project, however, indicated that currently there is generally no such link between robot development and policy/standard making, or that such a link does exist but that it is either too complex and lacks openness, is simply far too complex or only exists between robot developers and policy/standard makers, thereby failing to engage with other relevant stakeholders [17].

More specifically, in response to the question of whether a link between robot development and policymaking is currently missing, the responses were greatly mixed (See Fig. 2). While some believe that there is currently no such link between robot development and policy/standard making, others believe that such a link already exists but is too complex, too complex, and lacks openness or only exists between developers and standardization organizations. At the same time, only a very small minority believed that there is already such a link between robot development and policy/standard making [17].

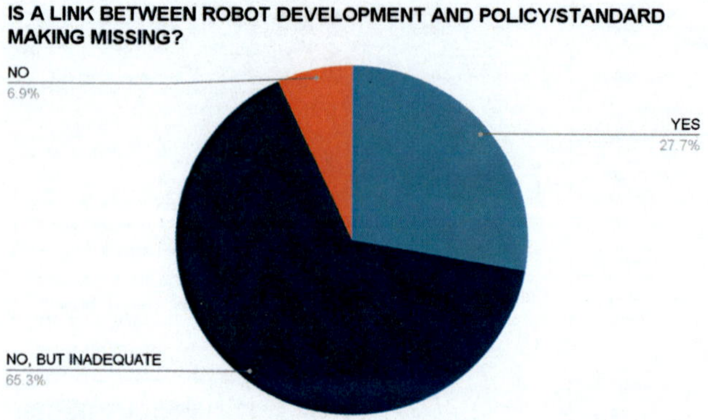

Fig. 2. Is a link between robot development and policy/standard making missing?

This all indicates that for the regulatory approach proposed through LIAISON to be valuable and effective, a diverse group of stakeholders – including robot developers, manufacturers, policymakers, standardization organizations, legal scholars, and ethicists – should be involved and a more diverse range of regulatory issues ought to be discussed, beyond mere physical safety [17]. For instance, engagement with DIH AgROBOfood has presented the need for robot developers to pay attention to ethical, legal, and many other issues to determine if a robot is sufficiently safe to operate in the wild. Moreover, exploratory meetings with standard makers clarified the value of LIAISON in liaising standardization activities and robot development [18]. In particular, the findings obtained throughout the LIAISON project have indicated a serious need for cooperation between central policymakers and standardization institutes (e.g., ISO, BSI, CENELEC), user group initiatives, and policy/standard makers.

Beyond the need for cooperation, the results flowing from the LIAISON Project also highlight an overall lack of legal comprehension among robot developers, particularly concerning experience with standards, knowledge about the difference between public and private policymaking, and experience in applying standards. Interestingly, the results point out that smaller and younger companies often lack the necessary knowledge and understanding concerning the applicable legal frameworks and the difference between private and public policymaking within this context.

As robot manufacturers deal with many different legal frameworks, including standards and regulations, compliance tools, such as the COVR Toolkit, may increase clarity. However, the COVR Toolkit leaves considerable room for desire. For instance, it does not distinguish between directives (i.e., a legislative act that sets out a goal that all EU Member States must achieve while it remains up to the individual Member States to devise their laws on how to reach these goals) and regulations (i.e., a binding legislative act which must be applied in its entirety across the EU). It also does not explain the relationship between standards (usually non-binding) and the law (usually binding). Even though such tools may clarify the applicable legal framework for robot developers, this reduces the complexity in robot legal compliance. With this in mind, improving compliance tools may clarify what binding legal frameworks robot developers ought to comply with and which recommended frameworks concretize abstract legal principles.

In this sense, a more personalized, dynamic, and multidirectional model of information provision would give developers a clear picture of which laws apply to their robot technology and facilitate engagement or feedback within the broader community. However, it is essential to note that the tool is not intended to provide legal advice, and those offering such tools may want to avoid any liability resulting from it.

At the same time, possible avenues for expansion concerning LIAISON have been identified. These relate to the active involvement in standardization organizations, focus on harmonization activities, and legal and educational participation in Digital Innovation Hubs to create more legal awareness among the involved communities of robot developers.

Bearing this all in mind, the LIAISON robot governance model could be highly beneficial in improving the communication between all stakeholders involved in the development and regulation. Over time, such engagement and knowledge generation could facilitate identifying dissonances and share lessons learned and tips from developers to prompt policymakers to enact policies that are more attuned to stakeholder needs and rights by allowing robot developers to make their voice heard in robot governance activities [17].

5 Conclusions

LIAISON has tested the theoretical model of a dynamic, iterative regulatory process for robot governance in practice, aiming to channel robot policy development from a bottom-up perspective towards a combined top-down/bottom-up model, leaving the door open for future modifications.

By clarifying what regulatory actions policymakers have to take to provide compliance guidance, explain unclear concepts or uncertain applicability domains to improve legal certainty, LIAISON has sought to understand what is necessary and what stakeholders should be involved to properly inform future regulatory developments for robot technology use and development at the European, National, Regional, or Municipal level.

To this end, LIAISON introduces a valuable model that could facilitate effective robot governance and improve the communication between all stakeholders involved in developing regulations (be it through public or private bodies) for new technologies.

Even more so, the importance and added value of combining this policy appraisal mechanism with compliance tools have already been confirmed in the literature, which stresses that 'better tools and appraisal design can lead directly to better policy appraisal and hence better policies' [19].

In this sense, the findings obtained and lessons learned throughout the LIAISON Project will complement the existing knowledge on the ethical, legal, and societal (ELS) challenges posed by robotics by providing clarity on how to address pressing but still uncovered safety challenges raised by robots, and represent a practical, valuable tool to advance social goals in an increasingly robotized world.

Acknowledgements. This contribution was written as part of the LIAISON project, an FSTP from the H2020 COVR project, which has received funding from the European Union's Horizon 2020 Research and innovation programme under grant agreement No 779966.

References

1. Marchant, G.E., Allenby, B.R., Herkert, J.R. (eds.): The Growing Gap Between Emerging Technologies and Legal-Ethical Oversight. Springer Netherlands, Dordrecht (2011). https://doi.org/10.1007/978-94-007-1356-7
2. Newlands, G., Lutz, C., Tamò-Larrieux, A., Fosch-Villaronga, E., Scheitlin, G., Harasgama, R.: Innovation under pressure: implications for data privacy during the Covid-19 pandemic. Big Data Soc. SAGE **7**(2), 1–14 (2020)
3. Holder, C., Khurana, V., Harrison, F., Jacobs, L.: Robotics and law: Key legal and regulatory implications of the robotics age (Part I of II). Comput. Law Secur. Rev. **32**(3), 383–402 (2016)
4. Sucha, V., Sienkiewicz, M.: Science for Policy Handbook. European Commission, Joint Research Centre (JRC). Elsevier (2020)
5. Fosch-Villaronga, E.: Robots, Healthcare, and the Law: Regulating Automation in Personal Care. 1 (edn.) 2020, Routledge, Milton (2019)
6. Gruber, K.: Is the future of medical diagnosis in computer algorithms? Lancet Digit. Health **1**(1), e15–e16 (2019)
7. Fosch-Villaronga, E., Heldeweg, M.: Regulation, I presume? Said the robot – towards an iterative regulatory process for robot governance. Comput. Law Secur. Rep. **34**(6), 1258–1277 (2018)
8. Spiliopoulou-Kaparia, M.: The evaluation of Directive 85/374/EEC on liability for defective products and Directive 2006/42/EC on machinery. In: Proceedings of the European Stakeholder Forum – Workshop on Regulatory Challenges for a Digitizing Industry. Essen (2017)
9. Marchant, G.E., Wallach, W.: Coordinating technology governance. Issues Sci. Technol. **31**(4), 43 (2015)
10. European Parliament resolution of 16 February 2017 with recommendations to the Commission on Civil Law Rules on Robotics (2015/2103(INL)) (2017)
11. A Bill to authorize an emerging technology policy lab within the General Services Administration, and for other purposes, S. 3502, 115th Cong. (2018). https://www.congress.gov/115/bills/s3502/BILLS-115s3502is.pdf
12. Leroux, P., Roy, S.: Designing and optimizing swarming in a distributed base station network: application to power control. ACM Trans. Auton. Adapt. Syst. (TAAS) **7**(2), 1–19 (2012)
13. Leenes, R., et al.: Regulatory challenges of robotics: some guidelines for addressing legal and ethical issues. Law Innovation Technol. **9**(1), 1–44 (2017)
14. Fosch-Villaronga, E., Golia, Jr. A.: Robots, standards and the law. Rivalries between private standards and public policymaking for robot governance. Comput. Law Secur. Rev. **35**(2), 129–144 (2019). https://doi.org/10.1016/j.clsr.2018.12.009
15. Fenwick, M., Kaal, W.A., Vermeulen, E.P.: Regulation tomorrow: what happens when technology is faster than the law. Am U Bus. L. Rev. **6**, 561 (2017)
16. Drukarch, H., Calleja, C., Fosch-Villaronga, E.: An iterative regulatory process for robot governance. Accepted for presentation at the DATA for Policy 2021 Conference, 14–16 September 2021. https://zenodo.org/record/5211103#.YfJLCljMK3K
17. Villaronga, E.F., Drukarch, H.: H2020 COVR FSTP LIAISON – D1.2 report on usefulness of LIAISON. eLaw Center for Law and Digital Technologies at Leiden University (2021). https://scholarlypublications.universiteitleiden.nl/handle/1887/3204019

18. Villaronga, E.F., Drukarch, H.: H2020 COVR FSTP LIAISON – D1.3 report on exploratory meetings with relevant policy/standard makers. eLaw Center for Law and Digital Technologies at Leiden University (2021). https://scholarlypublications.universiteitleiden.nl/handle/1887/3204021
19. Adelle, C., Jordan, A., Turnpenny, J.: Proceeding in parallel or drifting apart? a systematic review of policy appraisal research and practices. Environ. Plan. C: Gov. Policy **30**(3), 401–415 (2012). https://doi.org/10.1068/c11104

Towards Evidence-Based Standard-Making for Robot Governance

Carlos J. Calleja Ahmad(✉), H. Drukarch, and E. Fosch-Villaronga

eLaw Center for Law and Digital Technologies, Leiden University, Leiden, The Netherlands
{c.j.calleja.ahmad,h.g.drukarch,
e.fosch.villaronga}@law.leidenuniv.nl

Abstract. Despite the growing body of literature highlighting the legal and ethical questions robots raise, robot developers struggle to incorporate other aspects than mere physical safety into robot design to make them comprehensively safe. The chemical, food, and pharmaceutical industries established years ago use evidence-based frameworks that ensure the safety of these products EU-wide. However, these evidence-based frameworks have yet to be seen for robot technology. As a result, current robot technology raises many legal and ethical issues. The PROPELLING project aims to investigate how robot testbeds can be harnessed as data generators for standard-makers. To this end, the project focuses on testing safety requirements for lower-limb exoskeletons to understand whether standards, particularly ISO 13482:2014, address safety sufficiently and comprehensively and use the H2020 Eurobench testing beds and data as a means for appraising the standard. We suggest that linking experimentation settings with standard-making processes could speed up the creation, revision, or discontinuation of norms governing robot technology.

1 Introduction

Robotics and Artificial Intelligence (AI) have increased productivity and resource efficiency in the industrial and retail sectors, and now there is an emerging interest in realizing a comparable transformation in other sectors, such as healthcare. Such a shift is encouraged by the urge to increase care quality and safety while simultaneously restraining expenditure [1] and lately reducing human contact [2]. However, inserting robots in such a remarkably sensitive domain of application raises puzzling legal and ethical considerations [3, 4].

One of these considerations is the increasing disconnect between policy and technological progress [5], which is evident in robots used in healthcare applications. From exoskeletons to lightweight robotic suits, wearable robots are changing dynamically and rapidly, challenging the timeliness of laws and regulatory standards that were not prepared for robots that would help wheelchair users walk again [6]. These technologies

This contribution was written as part of the PROPELLING project, an FSTP from the H2020 Eurobench project, which has received funding from the European Union's Horizon 2020 Research and innovation programme under grant agreement No. 779963.

M. A. Grau Ruiz (Ed.): INBOTS 2021, BIOSYSROB 30, pp. 220–227, 2022.
https://doi.org/10.1007/978-3-031-04305-5_36

commonly raise ethical, legal, and social concerns related to the personal and psychological impact of such technologies on vulnerable individuals, like disabled people and their families [7].

In the specific case of robotic exoskeletons, the primary standard that governs them—ISO 13482:2014 Safety Requirements for Personal Care Robots [8]—seemingly fails to cover the hazards associated with these devices comprehensively [4]. There is no travel instability safeguard measures for lower-limb exoskeletons even though balance loss is the second cause of falling among older adults. Also, collisions with other objects are not considered to be a hazard. Most importantly, psychological hazards, such as fear of falling or anxiety, remain unexplored because of the traditional understanding that safety refers uniquely to physical matters [9]. The standard itself recognizes the lack of measures for specific categories of people, like children, pregnant women, and elderly people. However, it did not provide any specific guidance for them, failing to recognize the unique characteristics of individuals that are essential for the correct functioning of the devices [10].

It is in this context and based on the idea that information could help estimate risks and choose proper safeguards to ensure comprehensive safety that the Financial Support to Third Parties (FSTP) PROPELLING[1] financed by the H2020 Eurobench project[2] was born. We introduce the background and foreseen goals of PROPELLING in this contribution.

2 The FSTP Propelling within H2020 Eurobench

Propelling stands for "Pushing forward RObot develoPmEnt for LawmakING", and investigates how to use testing facilities to advance robot regulation.

A literature review reveals a paucity in harnessing research and development outcomes to improve existing regulatory instruments [11]. Indeed, while other industries such as the food or the chemistry industry use evidence-based models to aframeworks have yet to be seen for robotics. Instead, the regulatory landscape for these devices is currently populated by a myriad of fragmented regulations [12], abstract codes of conduct, and ethical guidelines that lack the necessary empirical grounding to adequately inform the practices of researchers, designers, and developers. Beyond the question of which regulations better account for user rights [13], these pieces were enacted when this technology did not yet exist. Moreover, to ensure that they are future-proof, these regulations are primarily technology-neutral.

An often-disregarded problem in this regard is the assumption that standards frame technology developments accurately [4, 14, 15]. There are some indications that ISO 13482:2014 fails to address safety comprehensively and accurately. For instance, while balance loss is the second cause of falls among older adults, travel instability measures do not apply to lower-limb exoskeletons (ISO 13482:2014 Annex A.1 Hazard item 59). Another example is that, while obstacles can pose a risk to normal gait (e.g., stairs, objects), ISO 13482:2014, however, states that collisions with safety-related objects,

[1] *See* https://www.universiteitleiden.nl/en/research/research-projects/law/propelling.
[2] *See* https://eurobench2020.eu/.

other robots, fragile safety-related objects, walls, permanent/unmovable barriers 'are not applicable to restraint-type physical assistant robots' (ISO 13482:2014 Annex A.1 Hazard item 62–63).

Such a limitation has one of its sources: the lack of methods for gathering and harnessing data for research and development. Indeed, a recently published technical report admits that testing methods for care robots "have not been implemented or evaluated broadly" [16]. Whereas these reflections often occur mainly in testing beds, where a prototype confronts the requirements of established standards [6, 11], they could also serve to improve regulations to establish new safety requirements, reformulate existing criteria, or abandon specific provisions.

Using an evidence-based approach to improve ISO 13482:2014 demands defining a methodology for collecting the necessary data. There are several ways to use evidence-based mechanisms, like collecting feedback from developers [17] or deploying robots in real-life situations in testing zones to gather information from their performance [11, 18]. At PROPELLING, we aim at investigating to what extent robot testbeds can be harnessed as data generators for standard-makers. To this end, we identify regulatory gaps and legal inconsistencies through a literature search and, based on the resulting findings, test particular safety requirements for lower-limb exoskeletons to understand whether standards and particularly ISO 13482:2014, address safety sufficiently and comprehensively.

3 Literature Search Findings

ISO 13482:2014 does not define 'personal care', whereas the definition of 'personal care robot' remains vague. While ISO 13482:2014 aimed to clarify the field, it created many new confusing categories. For instance, ISO 13482:2014 establishes a whole category, physical assistant robots, that can be used for rehabilitation purposes (with a clear medical intended purpose) or to augment physical capabilities to help workers in a factory lift heavier weights. In light of these two intended purposes, companies like Cyberdyne had two types of physical assistant robotic devices: those with medical device certification and those without certification. There are also robots for neurorehabilitation therapies like the robotic seal Paro marketed as medical devices in the U.S. and as toy robots in Europe [19].

Additionally, the theoretical categorization of robots by function can lead to confusion and under-regulation, where robots posing similar risks might fall outside the standard's scope. By having an unclear scope, robot developers can easily comply with the wrong instrument or put in place appropriate safeguards. That scenario not only exposes users to additional hazards but brings in the risk of sanctions and further responsibilities due to a lack of diligence.

Apart from an unclear scope, ISO 13482:2014 lacks specific requirements for different categories of people, like children and women [10]. It is thus uncertain whether devices designed in compliance with the standard are suitable for particular subjects and questions how and to what extent diversity and inclusion as concepts are considered. For instance, there are no guidelines on which design decisions roboticists should implement to ensure that male exoskeletons are fit not only for a standard male body but also for women's bodies (e.g., softer skin, wider hips).

We also identified room for improvement in what concerns the identification of hazards. ISO 13482:2014 provides partial coverage of close physical HRI tasks, particularly sit-to-stand support. However, it does not provide adequate guidelines on the level of support a user needs to move from sit-to-stand and stand-to-sit. Additionally, the standard is not suitable for ensuring safety in public spaces, nor does it consider whether touching wrong body parts could lead to safety hazards. Although robotic exoskeletons integrate software and hardware components, internal and external vulnerabilities remain ignored.

A crucial area in need of improvement is that of psychological hazards. How individuals interact with robots in a given context and for some time can have adverse short and long-term effects on their well-being and overall behavior, including how they perceive themselves [7, 9]. ISO 13482:2014 also recognizes that hazards can arise from physical and mental aspects of using the personal care robot. However, while the physical hazards are very prominently addressed in the standard, psychological aspects are only stress and discomfort derived from robot use. The standard thus ignores significant psychological hazards, like the fear of falling that a user might experience while wearing the exoskeleton for the first time after being in a wheelchair.

Regarding safeguard measures, ISO 13482:2014 lacks specific guidelines to deal with inevitable or existing falls [4]. Ideally, robots would be safe enough to render falls rare. Nonetheless, in the case of lower-limb exoskeletons, where the device is fastened to the user's body, the possibility of falls cannot be excluded. Without standardized strategies to deal with them, users might suffer injuries. Relatedly, the ISO standard does not provide parameters for activating protective stops. There is no method to determine protective stop space or the area(s) where the robot should or should not stop. ISO 13482:2014 only mentions that the robot should notify its status to the user if the battery falls below a certain threshold, and those that can trap humans in isolated locations shall include means of summoning assistance. However, the standard does not mention specific spaces where the device shall not provide support, like crowded areas or walking on stairs. Users might thus suffer additional injuries if the robot stops in a circumstance where support is essential for avoiding falls or impending collisions. Moreover, users might not feel in control of the device if they cannot stop because the button is unreachable for them or if others can reach the button and stop the exoskeleton without their consent.

ISO 134982:2014 also needs improvements with regards to the information for use. The standard mentions that information for use shall 'include the need for proper training to avoid operator travel time being longer than recommended' [8]. However, there are no requirements for information to third parties, such as the trainers, family members, or bystanders.

Lastly, there are no performance standards or test method procedures to obtain measurement data. Remarkably, there are no commonly agreed physiological measurements to provide data about the effects of long-term use of exoskeletons. In connection with that, ISO 13482:2014 lacks usable HRI models to perform model-based risk analysis and forecast the implications of close HRI.

4 Robot Regulatory Gap Testing

Propelling explores the idea to extract policy-relevant data from robot testbeds, where prototypes go through different tests, and the obtained data is then used to improve the device. The project departs from the premise that these kinds of settings could also be a source of evidence for improving regulations for robotic exoskeletons and, particularly, ISO 13482:2014. Among the myriad of hazards and safety safeguards that our literature search suggests are missing in the standard, PROPELLING focuses on the fear of falling (FoF), control algorithms for protective stops and graceful collapsing, and instability in a collision (see Table 1).

Table 1. Areas of focus for propelling

Area in need of regulatory improvement	Main section within ISO 13482:2014
Fear of falling	Section 5.9, Annex A.1, 48–55
Protective stops and graceful collapsing	Section 6.2
Instability in a collision	Section 5.10.5 Annex A.1, 62–63

These elements have in common that they remain unaddressed within the ISO 13482:2014 as specific hazards or as safety requirements. In this sense, they serve to appraise the possibilities of harnessing information from experimentation and development to improve existing standards. Furthermore, we experiment with a comprehensive set of volunteers (see Table 2) to determine whether specific standards demand data on different categories of people.

Table 2. Volunteers for propelling

User	Characteristics
1	Female, healthy, young adult (>35), weight: 65–80 kg, height: 165–180 cm
2	Male, healthy, adult (<45), weight: 70–90 kg, height: 170–190 cm
3	Male, patient, older adult (71), weight: 70–90 kg, height: 170–190 cm
4	Male, healthy, adult (>35), weight: 70–90 kg, height: 170–190 cm

Under PROPELLING, a multidisciplinary team will be executing five protocols during two weeks of experimentation to assess the performance of a lower-limb exoskeleton. For this purpose, the EUROBENCH test facilities, including software and databases, will be used to generate evidence-based knowledge set to improve regulatory standards and, particularly, ISO 13482:2014. In the following section, we provide an example concerning protective stops.

5 Improving the Knowledge about Protective Stops

Emergency stop, protective stop, and graceful collapsing mechanisms are systems to stop the performance of a task while safeguarding exoskeleton users (ISO 13482:2014 Section 3.17). ISO 13482:2014 obliges manufacturers to include these features if the robot operation needs to stop entirely or safely control the safeguarded hazard (ISO 13482:2014 Section 6.2). While these robot stopping functions are geared towards an 'interruption of operation that allows an orderly cessation of motion for safeguarding purposes' (ibid.), these robots work in close interaction with humans, and a cessation of the motion may entail risks to the user wearing or using the robot. The standard nonetheless falls short in providing more specific guidelines on how the initiation of such stop mechanisms may impact the user.

(1) First, there is no guidance on how reachable the button initiating a protective stop should be. ISO 13482:2014 does not clarify whether the button should be located where it is easily accessible to the user or in the rear part, like many devices, where other persons could access it. In this sense, van Herpen et al. (2019) report a case where a user's partner erroneously pushed the button, causing an intra-articular fracture of the left tibial plateau to the exoskeleton user (2019).
(2) Second, ISO does not include specific guidelines on scenarios when protective stops should be initiated or whether this feature should be automated (see ISO 13482:2014 Section 6.2.2.3). That leads to the risk of triggering a protective stop in the circumstances where halting the device would increase the risk of harm, e.g., when a user is about to sit down, and then the robot stops directly and leaves the person standing up. In a sense, there is no information on managing what happens after the protective stop is activated.
(3) Third, ISO fails to specify in which circumstances users can restart the device manually or whether an automatic restart is a better alternative. There is a reported case in which the immediate cause of a fracture was that the user activated the exoskeleton right after it collapsed [20]. Indeed, the device did not detect that her lower extremities were misaligned due to the previous collapsing. That case highlights the need to include guidelines on how to restart the device.

Thanks to EUROBENCH's facilities and software, PROPELLING has been able to specify:

- Where the button initiating protective stops should be located to improve the user sense of control [7],
- Instances in which protective stops should be automatically activated or when their activation puts greater danger to the user [6], and
- Guidelines for restarting the device once a protective stop has been triggered [6].

We learned from the experiments that when the device stops because of the activation of the emergency stop, power is cut, and the device collapses, which could endanger the user if activated while climbing stairs. Also, when the protective stop is activated via a tablet, the robot stops suddenly in the position where it is. In such instances, the user is left in very uncomfortable positions, and it is unclear how long they can wait until there is no subsequent harm.

6 Conclusions

Testing zones for robot technology show the potential to generate policy-relevant data that can ulteriorly inform policy change. Linking experimentation settings with standard-making processes could, in the future, speed the creation, revision, or discontinuation of norms governing robot technology, increase their effectiveness in ensuring overall safety, and the legal certainty regarding a fast-paced changing environment like robotics [4]. Still, evidence-based interventions face limitations when it comes to emerging robotics. Particularly, most of the testing methods for robotic exoskeletons have not been implemented nor assessed, thus hindering developers' application of safety standards. Furthermore, there is a need to develop testing methods for validating safety measures related to mental hazards.

On that basis, we proposed in PROPELLING using the Eurobench test facilities, including software and database, as data generators for evidence-based knowledge that will improve, instead of a specific robotic device, the requirements included in the standard itself. By identifying testing methods, we also provide insights into how developers can operationalize the standard to address robotic devices' safety and security risks.

In this contribution, we identify the need to revise ISO 13482:2014 to include novel hazards and safeguarding alternatives to cover risk steaming, such as psychological aspects, sex, and gender considerations [10]. This is particularly relevant considering that the standard had remained virtually untouched in the latest revision in 2020. Overall, we present evidence that there are valid reasons to believe policy-makers can harness that test benches to produce evidence-based interventions to regulate emerging robotics.

Acknowledgment. This contribution was written as part of the PROPELLING project, an FSTP from the H2020 Eurobench project, which has received funding from the European Union's Horizon 2020 Research and innovation programme under grant agreement No. 779963.

References

1. Alemzadeh, H., et al.: Adverse events in robotic surgery: a retrospective study of 14 years of FDA data. PLoS One **11**(4), e0151470–e0151470 (2016)
2. Aymerich-Franch, L., Ferrer, I.: The implementation of social robots during the COVID-19 pandemic (2020)
3. Vallor, S.: Carebots and caregivers: sustaining the ethical ideal of care in the twenty-first century. Philos. Technol. **24**(3), 251–268 (2011)
4. Fosch-Villaronga, E.: Robots, Healthcare, and the Law: Regulating Automation in Personal Care, 1st edn. Routledge, Milton (2019)
5. Sucha, V., Sienkiewicz, M.: Science for Policy Handbook. European Commission. Joint Research Centre (JRC). Elsevier (2020)

6. Calleja, C., Drukarch, H., Fosch-Villaronga, E.: Harnessing robot experimentation to optimise the regulatory framing of emerging robot technologies. In: Accepted for Presentation at the Data for Policy 2021 Conference, 14–16 September 2021. https://zenodo.org/record/5211110#.YRtaZS0Ro_V
7. Kapeller, A., Felzmann, H., Fosch-Villaronga, E., Hughes, A.M.: A taxonomy of ethical, legal and social implications of wearable robots: an expert perspective. Sci. Eng. Ethics 1–19 (2020). https://doi.org/10.1007/s11948-020-00268-4
8. ISO 13482:2014 Robots and robotic devices—Safety requirements for personal care robots, Standardization, Editor (2014)
9. Martinetti, A., Chemweno, P.K., Nizamis, K., Fosch-Villaronga, E.: Redefining safety in light of human-robot interaction: a critical review of current standards and regulations. Front. Chem. Eng. **3**(666237), 1–12 (2021). https://doi.org/10.3389/fceng.2021.666237
10. Søraa, R.A., Fosch-Villaronga, E.: Exoskeletons for all: the interplay between exoskeletons, inclusion, gender and intersectionality. Paladyn J. Behav. Rob. **11**(1), 217–227 (2020). https://doi.org/10.1515/pjbr-2020-0036
11. Fosch-Villaronga, E., Heldeweg, M.: "Regulation, I presume?" said the robot – Towards an iterative regulatory process for robot governance. Comput. Law Secur. Rep. **34**(6), 1258–1277 (2018)
12. Leenes, R., et al.: Regulatory challenges of robotics: some guidelines for addressing legal and ethical issues. Law Innov. Technol. **9**(1), 1–44 (2017)
13. Fosch-Villaronga, E., Golia Jr., A.: Robots, Standards and the Law. Rivalries between private standards and public policymaking for robot governance. Comput. Law Secur. Rev. **35**(2), 129–144 (2019). https://doi.org/10.1016/j.clsr.2018.12.009
14. Salvini, P., Paez-Granados, D., Billard, A.: On the safety of mobile robots serving in public spaces: identifying gaps in EN ISO 13482: 2014 and calling for a new standard. ACM Trans. Hum.-Rob. Interact. (THRI) **10**(3), 1–27 (2021)
15. Boada, J.P., Maestre, B.R., Genís, C.T.: The ethical issues of social assistive robotics: a critical literature review. Technol. Soc. **67**, 101726 (2021)
16. ISO/TR 23482-1:2020 Robotics—Application of ISO 13482—Part 1: Safety-related test methods. International Organization for Standardization (2020)
17. Fosch-Villaronga, E., Drukarch, H.: LIAISON D1.1 Report on safety standards uncovered challenges. Retrieved from eLaw - Center for Law and Digital Technologies, Leiden University (2021). https://hdl.handle.net/1887/3204017
18. Weng, Y.H., Sugahara, Y., Hashimoto, K., Takanishi, A.: Intersection of 'Tokku' special zone, robots, and the law: a case study on legal impacts to humanoid robots. Int. J. Soc. Robot. **7**(5), 841–857 (2015)
19. Fosch-Villaronga, E., Albo-Canals, J.: "I'll take care of you", said the robot. Paladyn J. Behav. Robot. **10**(1), 77–93 (2019)
20. Bessler, J., Prange-Lasonder, G.B., Schulte, R.V., Schaake, L., Prinsen, E.C., Buurke, J.H.: Occurrence and type of adverse events during the use of stationary gait robots—a systematic literature review. Front. Robot. AI **7**, 158 (2020)

Medical Nanorobots: The Biocompatibility Constraints of Clinical Translation

Saheli Datta Burton(✉)

Department of Science and Technology Studies, University College London, London, UK
saheli.burton@ucl.ac.uk

Abstract. Clinical translation remains the overarching aim of medical nanorobotics although little of the intensifying research in the area in the last two decades have so far been routinised into clinical practice. This research explores the underlying issues of biocompatibility as the primary constraint in the realization of these clinical translatability aims. I argue that this is in large part due to the weak integration of translation-related regulatory and medical-use considerations such as biocompatibility, toxicity etc. in nanorobotics research, especially in the upstream phases of research conceptualisation. Following on from this, the paper suggests meaningful inclusion of regulatory and clinical voices as key for improving translatability.

1 Introduction

Nanorobots encapsulate the unprecedented, complex and contingent conflation of artificial intelligence (AI) with embodiments at nanoscales. By autonomously traveling to hard-to-access sites within the human body, nanorobots are expected to deliver a variety of 'targeted' diagnostics and therapeutics from imaging and drug delivery, to editing cancer cells and surgical incisions. This, is the promise of nanorobotics, which in turn has unlocked millions in global research investment since the mid-90s to make it a reality [1]. At the same time, there is a growing acknowledgement that the realisation of this promise is unlikely without a well-defined consideration, characterisation and mitigation of the unprecedented risks that underlie a conflation of the human, the biological and the digital [10]. Thus, beyond the hype and hope of nanorobotics' promise for biomedicine and healthcare, the key question (as we explore here) is - *what will it take to translate nanorobots into clinical practice?*

As many have shown, the pathway for translating technology into clinical practice is untidy, difficult and narratives of "nonadoption, abandonment, and challenges to the scale- up, spread, and sustainability" are the norm than the exception [2]. Contemporary AI-based medical technologies such as clinical prediction models are no different. They face similar challenges of routinization with few adopted into clinical practice [3–5]. A key constraint is the lack of attention to and meaningful integration of clinician's

This work was supported by the UK Engineering and Physical Sciences Research Council [grant number EP/S035362/1] for the PETRAS project.

M. A. Grau Ruiz (Ed.): INBOTS 2021, BIOSYSROB 30, pp. 228–232, 2022.
https://doi.org/10.1007/978-3-031-04305-5_35

tacit knowledge, especially in the upstream phases of research and its conceptualisation [6]-[7], which remain overwhelmingly centered around 'engineering' disciplines like bioengineering [8] robotics [9, 10] and nanotechnology [3, 11]. As this brief paper will show, this is also the case for biointelligent systems, which center on engineering disciplines especially nanorobotic research for biomedical applications [3], where limited meaningful inclusion of clinical or regulatory voices [4] considerably weaken translatability.

2 Methods

Drawing on an exhaustive review of literature informed by discussions with basic scientists, policymakers, regulators and sociologists, this brief article contextualises the translatability implications of 'design-driven' aspirations towards higher *biointelligence* or *bio*-AI in emerging nanorobotic research. This is presented through a critical analysis of the unaddressed (bio)compatibility issues confronting the domain and its adverse implications for translatability. The study concludes with a reflection on the overarching interdisciplinarity problem that challenges attempts to realise the promise of nanorobotics-led targeted therapeutics.

Notably, the ethical, social and legal implications of nanorobots in particular, and the nanotechnology area more generally, are substantial and explored elsewhere but beyond the remit of this article [12–16].

3 Background

3.1 Aspirations of Higher AI

Nanorobotics, as the term suggests, brings together the domains of nanotechnology with bioengineering in robotics (biorobotics) and is inextricably rooted in the histories and hopes of each. From the rich history of efforts to design, build, operate and control autonomous mechanical devices since atleast the 12th century, nanorobotics today integrates the contemporary advances and hopes of AI to develop embodied intelligence (robots) that are increasingly responsive to its environment. In turn, the aspiration of robotics researchers, much like AI developers since the early 60s, became directed towards attaining an ever-higher level of robot intelligence aimed at the shared lofty goal of reaching (and possibly eventually surpassing) human intelligence. This aspiration remained unchanged even as robotics integrated with other co-emerging domains such as *biointelligence*, a co-developing biological offshoot of AI using molecular programming [17, 18] and DNA nanoscience [19] since the early 80s, eventually leading to the emergence of *nanorobotics* (robotics at micro- and nanoscales).

Today, nanorobotics is expected to bring about a new era of 'targeted' medicine through the development of targeted biomedical applications where nanorobots travel *in vivo* (inside the human body) to reach hard-to-access lesion sites inaccessible to current medicine. Likewise, research in biomedical nanorobots have expanded rapidly in the last two decades backed by immense global research and development funding [1, 20], although the realisation of nanorobotics' goals remains stubbornly elusive.

3.2 The Biocompatibility Problem

The problem is that realising the promise of micro- or nanorobot-led targeted therapies means grappling with issues of compatibility between *the biological* (the human *in vivo* environment) and *the bio-intelligent device.* In the medical nanorobotics domain, this first and foremost means grappling with the design-biocompatibility tensions around the central question of: '*how nanorobots will travel inside the human body to reach its target?*' to its target site inside the body.

So far, the approach to overcoming this challenge of 'how a nanorobot will reach its target' has overwhelmingly focused on 'designing' intelligent autonomous energy sources to actuate and provide navigational ability to nanorobots. Towards this end, nanorobotic engineers, driven primarily by aims of mimicking electronic computing's levels of higher intelligence using DNA nanoscience, have drawn on 'engineering' ontologies and epistemologies to achieve ever-higher levels of *biointelligence* for fabricating nanorobots. In turn, these aspirations towards higher bio-AI and the overwhelming attention to the 'design' calculus that has since dominated the logic of recent advances in the area [21–24].

Yet, this design-fabrication focus on energy sources has mostly come at the expense of attention to the issues of biocompatibility or toxicity along a nanorobot's *in vivo* journey to the target site. As a result, considerable resources have so far been devoted to studying a diverse range of organic (enzymes, sperms) and inorganic (magnetic, chemical) autonomous energy sources for nanorobots to reach its target, even though few of these 'energy sources' studied had defined biocompatibility profiles to begin with.

A key outcome of this attention to robot design instead of its compatibility with its environment is that the technological state-of-the-art in medical nanorobotics today reflects a limited consideration of the breadth of complexity that dynamic *in vivo* contexts involve. These contexts range from the complexities of the bloodstream through which nanorobots are expected to travel (pharmacokinetics) [25], degradation challenges in the systemic excretory phases to the long term impact on genetic or embryonic material etc. [26, 27]. As a result, translatability of medical nanorobotic candidates under development today or their usability in clinical settings, remains a long way off with only a few beginning to be tested on animal models.

4 Discussion

In turn, this weak design-biocompatibility engagement, mostly as a result of the uncritical quest for ever-higher levels of bio-AI at the expense of the nanorobot's adaptiveness to its *in vivo* environment, is symptomatic of a larger issue of weak interdisciplinary engagement. As such, studies reveal highly limited interdisciplinary collaboration across the constitutive disciplinary siloes of nanorobotics such as (bio/nano) engineering, medical practice and regulation [3, 28, 29, 30].Yet, the reality remains that without meaningful collaboration across these silos to ensure the safety and wellbeing of the human host (as well as that of future generations) even in the upstream research phases, realising the promise of nanorobotics is likely to remain challenging for the foreseeable future.

5 Conclusion

This article presented the implications of missed attention to meaningful considerations of *in vivo* contexts in the last two decades of intensifying nanorobotic research. I illuminated the translatability challenges as a result of the uncritical and unreflexive focus on design-driven 'technical' research for ever-higher levels of bio-AI with few if any qualitative commitment to or consideration of downstream clinical or regulatory components (see e.g., [25]).

The enduring reality for making nanorobotic translation realizable is that safety and wellbeing of the human host (the patient and the clinical setting) should have driven research especially in its upstream conceptualisation phases. For this to happen, the integration of translation-related regulatory and medical-use considerations such as biocompatibility, toxicity etc.are key and attainable through the meaningful inclusion of regulatory and medical expert voices especially in research conceptualization phases.

Acknowledgments. I would like to thank Professors Brian Salter and Nikolas Rose for their guidance and support as I developed this strand of research.

References

1. World Intellectual Property Office. Breakthrough Innovation and Economic Growth. Economics & Statistics Series, pp. 1–144 (2015). http://www.wipo.int/edocs/pubdocs/en/wipo_pub_944_2015.pdf
2. Greenhalgh, T., et al.: Beyond adoption: a new framework for theorizing and evaluating nonadoption, abandonment, and challenges to the scale-up, spread, and sustainability of health and care technologies. J. Med. Internet Research **19**(11), e367 (2017)
3. Burton, S.D.: Not anytime soon: the clinical translation of nanorobots and its biocompatibility-interdisciplinarity critique. In: Artificial Intelligence and Its Discontents, pp. 147–178. Palgrave Macmillan, Cham (2022)
4. Datta Burton, S., Mahfoud, T., Aicardi, C.C.R., Rose, N.: Clinical translation of computational brain models: understanding the salience of trust in clinician-researcher relationships. Special issue 'AI and its discontents'. Interdisciplinary Sci. Rev. **46**(1), 138–157 (2021)
5. Wessler, B.S., Paulus, J., Lundquist, C.M., Ajlan, M., Natto, Z., Janes, W.A., et al.: Tufts PACE clinical predictive model registry: update 1990 through 2015. Diag. Prognost. Res. **1**(1), 20 (2017)
6. Allegaert, K., Smits, A., van den Anker, J.N.: Physiologically based pharmacokinetic modeling in pediatric drug development: a clinician's request for a more integrated approach. J. Biomed. Biotech. 1–3 (2012). Article ID 103763https://doi.org/10.1155/2012/103763
7. Wyatt, J.C., Altman, D.G.: Commentary: prognostic models: clinically useful or quickly forgotten? BMJ **311**, 1539 (1995)
8. Yang, Q., Zimmerman, J., Steinfeld, A., Carey, L., Antaki, J.F.: Investigating the heart pump implant decision process. In: Proceedings of the 2016 CHI Conference on Human Factors in Computing Systems, pp. 4477–88. ACM, New York (2017)
9. Michalec, O., O'Donovan, C., Sobhani, M.: What is robotics made of? The interdisciplinary politics of robotics research. Hum. Social Sci. Commun. **8**(1), 65 (2021)
10. Patel, M., et al.: Harnessing interdisciplinarity to promote the ethical design of AI systems. In: ECIAIR 2019, European Conf. on the Impact of Artificial Intelligence & Robotics, Oxford, p. 246 (2019)

11. Contera, S.: Nano Comes to Life: How Nanotechnology is Transforming Medicine and the Future of Biology. Princeton University Press (2019)
12. Allhoff, F.: The coming era of nanomedicine. Am. J. Bioethics **9**(10), 3–11 (2009)
13. Allhoff, F., Lin, P., Moor, J.H., Weckert, J.: Nanoethics: The Ethical and Social Implications of Nanotechnology. Wiley, Hoboken (2007)
14. Dupuy, J.P.: Some pitfalls in the philosophical foundations of nanoethics. J. Med. Philo. **32**(3), 237–261 (2007)
15. Grunwald, A.: From speculative nanoethics to explorative philosophy of nanotechnology. NanoEthics **4**(2), 91–101 (2010)
16. Zainal Abidin, H.F., Hassan, K.H., Zainol, Z.A.: Regulating risk of nanomaterials for workers through soft law approach. NanoEthics **14**(2), 155–167 (2020). https://doi.org/10.1007/s11569-020-00363-7
17. Adleman, L.M.: Molecular computation of solutions to combinatorial problems. Science **266**(5187), 1021–1024 (1994)
18. Adleman, L.M.: Computing with DNA. Sci. Am. **279**(2), 54–61 (1998)
19. Seeman, N.C.: Nucleic acid junctions and lattices. J. Theor. Biol. **99**(2), 237–247 (1982)
20. Jia, L.: Global governmental investment in nanotechnologies. Curr. Nanosci. **1**(3), 263–266 (2005)
21. Elmowafy, E.M., Tiboni, M., Soliman, M.E.: Biocompatibility, biodegradation and biomedical applications of poly(lactic acid)/poly(Lactic-Co-Glycolic acid) micro and nanoparticles. J. Pharm. Invest. **49**(4), 347–380 (2019)
22. Tian, T., Xiao, D., Zhang, T., Li, Y., Shi, S., Zhong, W., et al.: A framework nucleic acid based robotic nanobee for active targeting therapy. Adv. Funct. Mat. **31**(5), 1–9 (2021)
23. Wu, Y., Wu, Z., Lin, X., He, Q., Li, J.: Autonomous movement of controllable assembled janus capsule motors. ACS Nano (12), 10910–10916 (2012)
24. Ye, Y., Gu, Z., Yao, C., Luo, D., Yang, D.: Nucleic acid-based functional nanomaterials as advanced cancer therapeutics. Small **15**(26), 1900172 (2019)
25. Saifi, M.A., Poduri, R., Godugu, C.: Nanomedicine: Implications of nanotechnology. In: Saifi, M.A., Poduri, R., Godugu, C. (eds.) Nanomedicine: Implications of Nanotoxicology. Drug Discovery and Development: From Targets and Molecules to Medicines, pp. 393–415 (2020)
26. Asare, N., et al.: Cytotoxic and genotoxic effects of silver nanoparticles in testicular cells. Toxicology **291**(1–3), 65–72 (2012)
27. Gatti, A., Montanari, M., Vaseashta, A.: Nanopathology - risk assessment of mysterious cryptogenic diseases. In: Vaseashta, A. (ed.) Life Cycle Analysis of Nanoparticles: Reducing Risk and Liability, pp. 143–171. DEStech Publications, Inc. (2015)
28. Patel, M., et al.: Harnessing interdisciplinarity to promote the ethical design of AI systems. In: ECIAIR 2019, European Conference on the Impact of Artificial Intelligence and Robotics, Oxford, UK, p. 246 (2019)
29. Sollini, M., Gelardi, F., Matassa, G., Delgado Bolton, R. C., Chiti, A., Kirienko, M.: Interdisciplinaridad: un requerimiento esencial para la traslación de investigación en radiómica a la práctica clínica. Rev. Española Medicina Nuclear Imagen Mol. **39**(3), 146–156 (2020)
30. Wong, M.C., Turner, P., Yee, K.C.: Involving clinicians in the development of an electronic clinical handover system-thinking systems not just technology. Stud. Health Tech. Inform. **136**, 490–495 (2008)

Check for updates

Regulation of Artificial Intelligence in the United States

Harper Pack(✉)

Department of Computer Science, Northwestern University, Evanston, USA
harper.pack@northwestern.edu

Abstract. Concerns about national competitiveness and social harmony motivate the development of artificial intelligence regulation (AI) in the United States (US). Regulations promote the development of technology standards, a national talent base, and autonomous vehicles and constrain the use of personal data, biased algorithms, and certain facial recognition technologies. The United States largely relies on preexisting agencies to administer and enforce these regulations; going forward, one can expect additional government entities to issue guidance on how they will approach regulating artificial intelligence technologies and applications that fall within the purview of their authority.

1 Introduction

This document provides a brief overview of regulation about and around AI in the US. It begins with an analysis of the objectives of US AI regulation, uses this as a frame for reviewing the extant types of enforcement and regulation, and then concludes with thoughts on the future of such regulation.

2 Why Does the us Regulate AI?

The US understands AI to be "…a transformative technology that holds promise for tremendous societal and economic benefit" [1]. Any nation-state would seek greater control over such a technology. The US, however, envisions itself as a leader of nation-states and this colors its approach to AI: in 2019, the President declared that "Continued American leadership in Artificial Intelligence is of paramount importance…" [2]. One can best understand the intentions of the US with respect to AI regulation through this lens: a vehicle for preserving and advancing its global pre-eminence.

On the one hand, the US seeks to promote AI for "…maintaining the economic and national security of the United States" [2]. On the other hand, the US seeks to restrict AI to "ensure that everyone has the opportunity to help in building an AI-enhanced society and to participate in its benefits." [1] The US wants to reap the geopolitical benefits of advanced AI capabilities while avoiding the social costs which could be wrought from any discriminatory, dangerous, or unfair applications of those capabilities.

Harper Pack is a Research Associate working with the C3 Lab at Northwestern University

M. A. Grau Ruiz (Ed.): INBOTS 2021, BIOSYSROB 30, pp. 233–237, 2022.
https://doi.org/10.1007/978-3-031-04305-5_38

3 What AI Does the us Regulate?

These contrasting objectives are, of course, inseparable. Like any technology (especially one encompassing such a wide variety of tools, methods, and components), the impact of AI depends on the nature of its use. The US regulatory scope reflects this reality: while there are some general policies for AI, there are also regulations governing its application to specific use cases.

3.1 Standards and Adoption

The federal government has encouraged the development of AI, particularly within governmental agencies. Some agencies, in particular the Federal Trade Commission (FTC), have found it within their purview to foster public dialogue on emerging AI technologies. The 2020 National AI Initiative Act (NAIIA) created a National AI Initiative Office (along with several committees) to facilitate a broader and more coordinated federal engagement with AI. The Act further directed the National Institute for Standards and Technology (NIST) to develop broad standards and frameworks for AI development, particularly with respect to risk management, trustworthiness, and bias (several agencies, such as the Department of Energy, received a comparable directive for best practices specific to their domain) [3].

There has not been a concerted focus at the state level, though some states have pursued similar ends. Alabama, New York, Texas, and Vermont have empowered groups to study, encourage, or advise on the adoption of AI [4]. California passed a resolution endorsing a set of AI Principles "as guiding values for the development of artificial intelligence and of related public policy", but these are not law [5].

3.2 Education and Research

The federal government has also sought to cultivate AI and related technological expertise and expand research into the same. In 2020, the National Science Foundation (NSF) established five national AI research institutes focused on specific AI topics, an effort further bolstered by the NAIIA [3, 6]. The NAIAA also seeks to increase the base of AI talent through "support for K-12 education and postsecondary educational programs, including workforce training and career and technical education programs" [3]. A bill currently under debate in the Senate, the Endless Frontier Act, could significantly expand these activities [7].

This has not yet been a major focus for state governments. Mississippi and Utah have each enacted policy to incorporate AI and technology concepts into public education [4].

3.3 Personal Information

The federal government has long regulated the collection and disclosure of personal information. Over time, individual pieces of legislation have arisen to provide protections for privacy and security of personal information across a variety of contexts (e.g., healthcare, financial services, surveillance, and media consumption) [8].

As such, there is no blanket federal data regulation. In contrast, states have all enacted regulations concerning personal information writ large. While details vary by state, their broader intentions are comparable to those behind the federal regulations: limiting the unwanted dissemination of personal information through requirements for information security and disclosure [8].

Illinois, California, and Virginia stand out among states in this regard. Illinois' Biometric Information Privacy Act defines biometric data as a special category of personal information which can only be collected pursuant to notifying and receiving consent from a person [9]. The California Consumer Privacy Act empowers consumers to obtain information about and (to varying extents based on the data and the collector) limit the collection and use of their personal data [10]. In March 2021, Virginia passed a similar law [11].

3.4 Algorithms and Automated Decision-Making

The federal government has also long regulated automated decision-making. The FTC Act governs unfair business practices which are deceptive or unreasonably damaging to customers [12]. Though it has long focused on algorithms in the context of credit, recent FTC guidance indicates that any algorithm which surveils individuals, exhibits material bias, or obscures the purpose or mechanics of its operation may be considered unfair [12]. There are not yet other major federal or state regulations on the use of algorithms.

3.5 Context-and Domain-Specific Policy

There are a growing number of use-case and domain- specific regulations for AI technology. These regulations are characterized by their common focus on the application, rather than the development, of AI. Broadly speaking, one can observe a similar pattern of regulation supporting AI which pertains to national economic growth and competitiveness and restricting AI which may offend social norms.

For example, many states passed legislation – with varying degrees of permissibility – to promote autonomous vehicles; national agencies such as the National Highway and Transportation Safety Administration (NHTSA) and the Federal Aviation Administration (FAA) have in turn issued guidance for autonomous machines [13].

There have also been regulations created to constrain the use of image and video analytics. Most notably, several states and municipalities have passed regulation to limit the use of facial recognition, particularly by law enforcement [14]. Illinois has passed a narrower law concerning the use of AI technology in hiring contexts, requiring employers obtain the consent of candidates before using AI interviewing tools [14]. Maryland and New York City have followed Illinois' lead in enacting similar measures which prohibit the non-consensual use of AI technologies within the hiring process [15].

4 How Does the US Administer its AI Regulation?

The US has created several task forces and committees whose sole purpose is to support and study the development of AI and its integration into commerce, bureaucracy, and society, such as the National AI Initiative Office created by the NAIIA [3]. The US does not, though, have a single authority for regulating AI. Instead, the US relies on existing agencies to supervise AI within their regulatory purview. For example, the US Food and Drug Administration (FDA) has a mandate for protecting public health with respect to the manufacture and consumption of food, medicine, and related products; it has no immediate relation to AI. However, as AI technologies evolve and encroach upon the public health domain, they become subject to the FDA's regulatory scope. Indeed, in 2021 the FDA released its "Artificial Intelligence and Machine Learning (AI/ML) Software as a Medical Device Action Plan" [16]. Other agencies, such as the aforementioned FTC, NSF, NIST, NHTSA, and FAA, have followed a similar approach in engaging with AI at its intersection with their primary organizational mandate.

The US does not solely rely on domain-specific institutions to administer and enforce its AI policy. For those rules which do not fall under the purview of existing authoritative body, the legal system effectively serves as the regulatory mechanism. The Illinois Biometric Information Privacy Act, for example, stipulates that "any person aggrieved by a violation of this Act shall have a right of action in a State circuit court…" and specifies damages which they can seek [17]. This private right of action has engendered dozens of class action lawsuits against companies alleged to have violated the Act [18].

5 What is the Future of US AI Regulation?

It is obviously difficult to predict the future of AI regulation in the US, not least due to the dynamic nature of the underlying technologies and the uncertainty inherent in the political landscape. Nonetheless, one might expect that as new use cases for AI emerge, the relevant domain authorities will develop appropriate oversight mechanisms. Further, states and municipalities may outpace the federal government in developing and implementing legislation for AI as it pertains to business and society; the federal government may subsequently explore adapting popular and effective components from these after observing their effects in practice.

6 Conclusion

The US seeks to harness the commercial and geopolitical benefits of leading the world in the development of AI technology while minimizing the risks inherent in the adoption of disruptive technologies. US AI regulations largely reflect this dual objective. The existing regulatory mandates of federal agencies have proven sufficiently broad to cover current AI use cases. As one looks ahead, one can expect AI to become increasingly integrated into society and in turn inspire additional sector-specific regulation from the relevant governmental agencies.

References

1. Networking and Information Technology Research and Development Subcommittee, et al.: The National Artificial Intelligence Research and Development Strategic Plan, Executive Office of the President of the United States, 13 October

2016. https://obamawhitehouse.archives.gov/sites/default/files/whitehouse_files/microsites/ostp/NSTC/national_ai_rd_strategic_plan.pdf
2. Executive Office of the President of the United States. AI for the American People, January 2021. https://trumpwhitehouse.archives.gov/ai/
3. H.R. 6395—116th Congress: National Defense Authorization Act for Fiscal Year 2021 (2020). https://www.govtrack.us/congress/bills/116/hr6395. Accessed 18 May 2021
4. Greenberg, P.: Legislation Related to Artificial Intelligence, NCSL, April 2021. www.ncsl.org/research/telecommunications-and-information-technology/2020-legislation-related-to-artificial-intelligence.aspx
5. State of California Endorses Asilomar AI Principles. Medium, Future of Life Institute, 31 August 2018, medium.com/@FLIxrisk/state- of-california-endorses-asilomar-ai-principles-69da95fb10e0
6. NSF Advances Artificial Intelligence Research with New Nationwide Institutes. NSF, 26 August 2020. www.nsf.gov/news/special_reports/announcements/082620.jsp
7. Shepardson, D., Michael, M.: U.S. senate panel approves tech bill to address China. Reuters, Thomson Reuters, 12 May 2021. https://www.reuters.com/world/us/us-senate-panel-vote-china-tech-bill-wednesday-2021-05-12
8. Chabinsky, S., Pittman, P.F.: Data protection 2020: laws and regulations: USA: ICLG. international comparative legal guides international business reports, global legal group 2 (2020). https://iclg.com/practice-areas/data-protection-laws-and-regulations/usa
9. Blaney, R., et al.: Litigation breeding ground: Illinois' Biometric Information Privacy Act. The National Law Review, 18 March 2021. www.natlawreview.com/article/litigation-breeding-ground-illinois-biometric-information-privacy-act
10. Bonta, R.: California Consumer Privacy Act (CCPA). Office of the Attorney General, 3 March 2021. https://oag.ca.gov/privacy/ccpa
11. Southwell, A.H., et al.: Virginia passes comprehensive privacy law. Gibson Dunn, 8 March 2021. www.gibsondunn.com/virginia-passes-comprehensive-privacy-law/
12. Jillson, E.: Aiming for truth, fairness, and equity in your company's use of AI. Federal Trade Commission, 19 April 2021. www.ftc.gov/news-events/blogs/business-blog/2021/04/aiming-truth-fairness-equity-your-companys-use-ai
13. Shinkle, D.: Autonomous Vehicles: Self-Driving Vehicles Enacted Legislation. NCSL, 18 February 2020. www.ncsl.org/research/transportation/autonomous-vehicles-self-driving-vehicles-enacted-legislation.aspx
14. Chae, Y.: U.S. AI regulation guide: legislative overview and practical considerations. RAIL J. Robot. Artif. Intell. Law, **3**(1), 17–40 (2020). www.bakermckenzie.com/-/media/files/people/chae-yoon/rail-us-ai-regulation-guide.pdf
15. Gajda, K.: New York City regulates workplace artificial intelligence recruitment and selection tools. The National Law Review, 18 November 2021. https://www.natlawreview.com/article/new-york-city-regulates-workplace-artificial-intelligence-recruitment-and-selection
16. Center for Devices and Radiological Health. Artificial Intelligence and Machine Learning in Software. U.S. Food and Drug Administration, FDA, 22 September 2021. https://www.fda.gov/medical-devices/software-medical-device-samd/artificial-intelligence-and-machine-learning-software-medical-device
17. The Biometric Information Privacy Act. (Source: P.A. 95–994, eff. 10–3–08.). https://www.ilga.gov/legislation/ilcs/ilcs3.asp?ActID=3004&ChapterID=57
18. Steven, G., Shinabarger, E.: Biometric privacy litigation: the next class action battleground. Bloomberg Law, Bloomberg, 9 January 2018. https://news.bloomberglaw.com/business-and-practice/biometric-privacy-litigation-the-next-class-action-battleground/

Priorities in the European Union's Digital Decade with Special Reference to the Spain Plan

Alfredo Luis Díaz Peña(✉)

Spanish Ministry of Finance, A Coruña, Spain
diazpalf@hotmail.com

Abstract. The "digital compass 2030" sets out the path to achieve the digital transformation of the European Union, whose objectives include the digitisation of companies, which integrates automation, Artificial Intelligence, Big Data and robotisation. We will briefly analyse the instruments to achieve this objective, especially the financial resources provided by the Recovery and Resilience Fund and, in particular, the Spain Plan. And finally, the repercussions of the digital economy from a fiscal point of view and the approaches to its treatment.

1 Introduction

At the European Union (EU) level, the Communication *"The Digital Compass 2030: Europe's Road to the Digital Decade (June 2021)"*, establishes the vision, objectives and ways to achieve a digital transformation by 2030. To this end, its achievement is linked to objectives, such as four cardinal points, hence the name compass, which are: digital skills, digital infrastructure, digitisation of businesses and public services.

The EU aims to be digitally sovereign in an open and interconnected world and to implement digital policies that empower people and businesses to take advantage of a more prosperous, sustainable and people-centred digital future. And, it is doing so[1], because of its importance for the recovery, prosperity, security and competitiveness of the Union and for the well-being of our societies.

It is not only about achieving this digitisation goal, but also about doing so in line with EU values, i.e., strengthening digital leadership and promoting people-centred, inclusive and sustainable policies. Bringing all these criteria together, the so-called "digital principles" were drawn up, at the service of all European citizens, which seek to put people at the heart of the digital transformation; solidarity and inclusion; freedom of choice; participation in the digital public space; safety; security; and sustainability.

To this end, the Commission is empowered to expand the range of policy instruments available to the European Union for the digital transformation, and to use all available instruments offered by industry, trade, competition, skills and education policies and

Deputy General Director for Payment Requests of the Recovery and Resilience Mechanism, Secretary General for European Funds, Directorate General for the Recovery and Resilience Plan and Mechanism

[1] European Council, conclusions of 25 March 2021.

M. A. Grau Ruiz (Ed.): INBOTS 2021, BIOSYSROB 30, pp. 238–246, 2022.
https://doi.org/10.1007/978-3-031-04305-5_39

research and innovation policies, as well as long-term funding instruments, to facilitate the digital transformation.

In the economic sphere, in the context of the digitisation of companies, it is stressed that, beyond new manufacturing and business processes in the context of a data economy, no company can be left behind, especially small companies, as they are the backbone of the European economy Therefore, the objective is that by 2030[2], at least 75% of European companies should reach cloud computing, big data and artificial intelligence services; and more than 90% of European SMEs should reach at least a basic level of digital intensity.

All these objectives, framed in the "digital compass", do not conflict with other EU policy objectives, but are coherent with them, as well as with EU budgetary instruments such as the cohesion programmes, the technical support instrument, the Digital Europe Programme, Horizon Europe and InvestEU, as well as with the Recovery and Resilience Plan whose Regulation (EU) 2021/241 sets a minimum allocation of 20% for the digital switchover.

2 Instrumentalisation Towards Digital Transformation

Which is the path to attain the goals framed in the "Digital Decade"[3] Firstly, achieving the expansion of high-performance broadband access for all citizens, including in rural and remote areas, is a key factor from an infrastructure point of view, as it has an impact not only on the quality and speed of connection but also on the problem of rural depopulation. Additionally, the further development of digital skills and competences of the workforce for full participation in the digital economy, in line with the digitisation of businesses and public services to achieve more efficient and user-friendly services. But all this must take place in the framework of a secure cyberspace, building greater trust, preserving freedom of expression and information, including media freedom and pluralism, and all in line with the ecological transition[4]. Ultimately, the aim will be to harness the power of technology to take climate action, protect biodiversity and restore nature, in line with the European Green Pact, the UN Sustainable Development Goals and the Paris Agreement.

Among the EU budget instruments, we have to pay particular attention to the Recovery and Resilience Plan. It foresees a large number of reforms and investments, some of which will involve technology, infrastructure, skills and digital processes to improve the EU's global competitiveness and thereby contribute to a more resilient, more innovative and less dependent EU by diversifying key supply chains.

---

2 https://eur-lex.europa.eu/resource.html?uri=cellar:6785f365-1627-11ec-b4fe-01aa75ed71a1.0002.02/DOC_1&format=PDF.

3 Communication from the Commission to the European Parliament, the Council, the European Economic and Social Committee and the Committee of the Regions. Establishment of a European declaration on digital rights and principles for the Digital Decade on 26 January 2022.file://C:/Users/KG000504/Downloads/COM_2022_27_1_EN_ACT_part1_LA380eaVUqfmgJ2s6UaeD7Fcbug_82699.pdf.

4 https://www.nature.com/articles/s42256-021-00436-6.

The Regulation (EU) 2021/241, in its recital 26 and in various provisions, states that the measures supported by the Facility, and included in the Recovery and Resilience Plans of each Member State, should represent an amount of at least 20% of the allocation of the Recovery and Resilience plan for digital expenditure. The fields and types of intervention included in Annex VII of this Regulation, with regard to the Methodology for digital labelling within the framework of the Mechanism, the development of highly specialised support services and facilities for public administrations and companies (High-Performance Computing HPC), (competence centres, cyber centres, testing and experimentation facilities in artificial intelligence, blockchain, internet of things, etc.) should be highlighted, with a coefficient of 100% for the calculation of aid for the digital transition. Among the Recovery and Resilience Plans presented by EU member states is the Spain Plan.

3 The Spain Plan

The Spain Plan, approved in June 2021[5], is a country project that incorporates an important agenda of investments and structural reforms, which are interrelated to achieve four cross-cutting objectives: to move towards a greener, more digital, more socially and territorially cohesive and more egalitarian Spain. These four axes are the ones that have guided the ten policy levers within which 30 components -and, therefore, all the reforms and investments of this Plan, are agglutinated.

In the area of policies aimed at digital transformation, and for an amount representing 28.2% of the total allocation of the Plan[6] calculated in accordance with the methodology of Annex VII of Regulation (EU) 2021/241 (approximately 19.606.89 million €) above the established minimum threshold; it integrates a set of measures aimed at an Administration for the twenty-first century (Lever Policy IV), for the modernisation and digitalisation of the industrial fabric and SMEs, recovery of tourism and boosting an entrepreneurial Spain (Lever Policy V), science and innovation as well as reinforcement of the capacities of the National Health System (Lever Policy VI) and, last but not least, education and knowledge, continuous training and skills development (Lever Policy VII).

With this important economic contribution, according to the Council's proposed document[7], Spain is expected to invest in areas such as digital skills, the digitisation of companies and the integration of advanced technologies in the business sector. From a horizontal point of view, these are expected to benefit the economy and the population as a whole (improved connectivity), complemented by the benefits for certain sectors such as tourism, SMEs, culture and the media, or specific segments of the population such as students, the unemployed and civil servants.

---

5 Proposal for an Implementing Decision (PD): https://ec.europa.eu/info/sites/default/files/com_322_1_es.pdf Council on the Approval of the Evaluation of the Spanish Recovery and Resilience Plan. https://planderecuperacion.gob.es/.

6 Recital 36 PS:https://ec.europa.eu/info/sites/default/files/com_322_1_es.pdf.

7 Recital 37 PD:. https://ec.europa.eu/info/sites/default/files/com_322_1_es.pdf.

To this end, the Plan includes seven strategic plans[8] within the framework of the Digital Agenda for Spain 2025, which today can be understood in the Digital Compass: the Plan for Connectivity, the Strategy to promote 5G technology, the National Artificial Intelligence Strategy, the National Digital Skills Plan, the Public Administration Digitalisation Plan, the SME Digitalisation Plan and the "Spain, Audiovisual Hub of Europe" Plan. Future robotics developments will probably have to accommodate to this changing context.

4 Measures for the Modernisation and Digitisation of the Industrial Fabric and SMES

The measures, reforms and investments aimed at meeting the challenge of digitalisation, as well as the robotisation of the Spanish economy, are structured around ten strategic axes:

- Digital connectivity.
- Continue to lead the Deployment of 5G technology.
- Strengthen the digital skills of workers and citizens.
- Strengthen Spain's cyber-security capacity.
- Digitise Public Administrations.
- Accelerate the digitalisation of companies, with a special focus on micro-SMEs and start-ups.
- Accelerate the digitalisation of the production model through digital transformation projects in strategic economic sectors, such as agri-food, mobility, health, tourism, trade and energy, among others.
- Improve Spain's attractiveness as a European audiovisual platform.
- Move towards a data economy, guaranteeing security and privacy and taking advantage of the opportunities offered by Artificial Intelligence with the aim that at least 25% of companies will be using Artificial Intelligence and Big Data within five years.

  Ensure rights in the new digital environment, in particular labor, consumer, citizens' and business rights.

All counting on a combination of public–private investment. On the one hand, the Next Generation (Recovery Plan) funds, with an investment of approximately €15 billion. On the other hand, with approximately €50 billion from the private sector. This is expected to contribute, transversally or vertically, to re-launch economic growth, increase productivity and, in parallel, reduce inequality through the exploitation of all the opportunities offered by new technologies.

In particular, the measures, investments and reforms that are financed by the Next Generation EU funds are integrated within the Lever V Policy of modernisation and digitisation of the industrial fabric and SMEs, recovery of tourism and promotion of

[8] The planned digital measures take into account the EU Digital Strategy, the European Industrial Strategy, the EU Skills Agenda, the European Pillar of Social Rights, the European Education Area and the "Spain, Audiovisual Hub of Europe" Plan.

an entrepreneurial nation Spain, and in particular within Components 12 to 15[9]. In the Component 12 "Industrial Policy Spain 2030", Investment 1 there are actions aimed at Sectoral data spaces (contribution to driving projects for the digitisation of strategic productive sectors) with a cost of 400 million €. The aim is to set up large common industrial and insurance data spaces to try to boost, through economies of scale, business innovation in sectors such as health and commerce. An example of this, according to the component's own document[10], is the development and sharing of common components (building blocks) and common infrastructures and data[11], such as cloud platforms and supercomputing and storage networks, massive data analysis and services and Artificial Intelligence. The use of these large spaces will make it possible to tackle, in better conditions, digitalisation and automation to optimise costs, quality and efficiency in the production processes of data collection and sensors, and in the actuator and robotic processes, resulting in an increase in the competitiveness of the associated industry.

In the case of Component 13 (C13) "Boosting SMEs"[12], the reforms and investments that make up this component are of a fundamental nature due to the role that small and medium-sized enterprises play in the European Union's economy. The document on *"A strategy for SMEs for a sustainable and digital Europe" (COM(2020) 103 final*[13]) states that more than 99% of EU enterprises are SMEs, employing some 100 million people and generating more than 50% of the EU's GDP. Therefore, their role is key in providing added value to all sectors of the economy and in the case of Spain it is even greater, five percentage points above European averages, while the size of companies is also smaller[14].

In this context, the C13 was designed so that the reforms and investments integrated in it would facilitate the creation, growth and restructuring of enterprises, improve the business climate and boost productivity growth through digitalisation, innovation and internationalisation. To achieve this objective, via digitalisation, two important driving conditions are necessary: a) the digital skills of workers; and b) the existence of clusters of cooperation and support between SMEs themselves, which allow the multiplier effects of networking and digitalisation to be exploited. These clusters are the subject of the current network of Innovative Business Clusters (AEI).

---

9 C12. Industrial Policy Spain 2030; C13. Promotion of SMEs; C14. Plan for the modernisation and competitiveness of the tourism sector and C15. Digital connectivity, boosting cybersecurity and deployment of 5G.

10 Component 12: https://www.lamoncloa.gob.es/temas/fondos-recuperacion/Documents/16062021-Componente12.pdf.

11 A European Strategy for Data: Big Data use by companies in Spain and Europe published in April 2020, the percentage of companies using these techniques is 11% in Spain compared to 12% of the European average, although it is below the percentage of use in countries such as France (16%), Germany (15%) and the United Kingdom (15%).

12 https://www.lamoncloa.gob.es/temas/fondos-recuperacion/Documents/16062021-Componente13.pdf.

13 https://eur-lex.europa.eu/legal-content/ES/TXT/?uri=CELEX:52020DC0103.

14 Eurostat, Spanish companies, including the self-employed, have an average workforce of 4.4 employees (5.9 on average in the EU) and less than half the average size of those in Germany (11.8) or the UK (9.4).

Although the investment in the digitalisation of SMEs is mainly contemplated under Investment 3 of C13, with a cost of €3,547.986 million, it is also complemented by other investments such as the connectivity vouchers in Component 15, or investment in the digital training of workers, entrepreneurs and agents in charge of promoting modernisation according to Component 19 (C19). An example of this complementarity can be seen, within the actions of Investment 3, in the "Agent of Change Programme"[15], which will subsidise, for up to one year and up to a maximum of 20,000 euros per company/agent, the costs of incorporating a professional in digital transformation (the so-called agent of change) for small and medium-sized companies. However, the training of these agents of change will be financed with the digital training investments of C19.

The importance of the tourism sector in Spain, with 12.3% of the GDP (in 2019) and a number of approximately 83 million foreign tourists received, led to the development of a specific Component 14 (C14) "Plan for the modernisation and competitiveness of the tourism sector"[16]. The investments aimed at digitalisation and intelligence for destinations and the tourism sector included in Investment 2 of C14, for an amount of 337 million €, deserve a special mention.

Digitalisation in this sector is strategic due to the type of tourist, more hyper-connected and demanding stays with greater added value. In fact this is one of the areas where the margin for improvement is widest and with the greatest potential impact for Artificial Intelligence (128% increase in the added value of the activity thanks to the application of AI).

The Component 15 "Digital connectivity, boosting cybersecurity and 5G deployment" is cross-cutting in nature, as it concentrates all the reforms and investments related to digital connectivity and 5G made under the Plan. Digital connectivity, access to networks, is a crucial factor, among others, for the development of economic activity, for increasing productivity and for boosting innovation.

Obviously, AI and Big Data due have a high incidence in our daily activities (with Internet search engines, e-commerce, personal assistants, etc.), but from a business and professional point of view, the aim is to optimise resources and automation, making it possible to predict demand and thus better manage logistical processes; to enable financial institutions to calculate their risks or predict their cash flows with greater precision, in short, their financing needs and solvency; or, in the field of education, to adapt learning, for example.

The Component 16 (C16) "National AI Strategy"[17] integrates, with an amount of €500 million, a set of investments aimed at boosting scientific research, technological development and innovation in AI as well as integrating AI into value chains to transform the economic fabric. The high-level goal of these investments is to increase the number of Spanish companies using AI from less than 15% at present to 25% by 2025.

---

[15] The "Agent of Change Programme" aims to support at least 15,000 small and medium-sized enterprises (10 to 249 employees) in their transformation process, which represents approximately 8.7% of the 171,884 small and medium-sized enterprises in Spain.

[16] https://www.lamoncloa.gob.es/temas/fondos-recuperacion/Documents/16062021-Componente14.pdf.

[17] https://planderecuperacion.gob.es/politicas-y-componentes/componente-16-estrategia-nacional-de-inteligencia-artificial.

Under C16, large projects that use AI and foster collaboration between research organisations, large companies and SMEs will be funded through grants. The aim, through large projects, is to achieve synergy between leading companies in the sector and small companies located throughout the territory, as well as to encourage projects that enable the creation of start-ups and new companies throughout the territory around the use of AI in priority sectors such as agri-food, tourism, energy, mobility, climate change, biomedicine and health.

Also, part of the investment will go towards attracting national or international talent in AI digital skills, financing the creation of university chairs whose activity is focused on research, dissemination and training, or aid programmes for SMEs to integrate AI and robotisation into their value chains.

The digital transformation requires that digital skills become basic tools to be able to take advantage of the opportunities it offers from an economic, social and environmental point of view. This is where Component 19 "National Digital Skills Plan" comes into play[18], targeting the entire population, from the digitisation of school to university, including upskilling and reskilling at work, with a special focus on closing the gender gap and boosting training in areas of demographic decline.

This plan is organised into four axes, with seven lines of action, which are expected to reach 80% of the Spanish population. These actions will cover, in a cross-cutting area, digital training for citizens (with emphasis on groups at risk of digital exclusion), and the fight against the gender digital divide. In the field of education, the development of digital skills for learning. Regarding employment, the training in digital skills for unemployed and employed people in the private sector, with special emphasis on unemployed young people, or staff in public administrations, as well as in small and medium enterprises (SMEs). Finally, encouraging digital professionals through the promotion of Information and Communication Technologies specialists.

In total, this component, from the perspective of the labor market -the active population, with four investments and a cost of 3,593.00 million €, aims to up-skilling and re-skilling. It also includes training in digital skills for SMEs, such as AI and cybersecurity for different areas, from the general public to more specific profiles.

5 Fiscal Impact of Digitisation from the Perspective of the Recovery Plan

The Component 28 (C28) "Adapting the tax system to the reality of the twenty-first century"[19], aims to contribute to Spain's economic recovery through growth, job creation and improved competitiveness. This will require a more equitable, progressive and fairer tax system in which the design of green taxation is further developed, the gender perspective is included and public policies of general interest are promoted. But also, to achieve a contribution from the digital economy, the Tax on Certain Digital Services

18 https://www.lamoncloa.gob.es/temas/fondos-recuperacion/Documents/16062021-Componente19.pdf.

19 https://www.lamoncloa.gob.es/temas/fondos-recuperacion/Documents/16062021-Componente28.pdf.

and some measures associated with the digital transformation are envisaged in Reform 5 of C28.

The digitalisation of the economy, and hence the new way of doing business, poses challenges from a tax perspective. This new way is largely underpinned by the ability to conduct business remotely, including across borders, with little or no physical presence, the importance of intangible assets, and the value of data and end-users' contributions to value creation.

In contrast, the current international tax framework is based above all on physical presence and is not yet designed to deal with new business models where companies can provide digital services in a country without being physically present there. These facts make it difficult to prevent the relocation of intangible assets to jurisdictions with little or no taxation. The component document recognises that this leads to a disconnect between the place where value is generated and the place where companies are taxed.

This has been a matter of international concern for years as evidenced by documents at the Organisation for Economic Cooperation and Development (OECD) and G20[20] level, or at EU level by the European Commission's Communication "A fair and effective EU taxation system for the Digital Single Market" adopted on 21 September 2017[21], and the package of proposals for Directives and Recommendation to achieve fair and effective taxation of the digital economy presented on 21 March 2018. Some multilateral steps are being taken.

6 Conclusion

Digital transformation is a key factor for completing the transition to a climate-neutral, circular and resilient economy as well as for the recovery, prosperity, security and competitiveness of the Union and increasing the well-being of our societies. Therefore, as the Council has pointed out: "in tomorrow's world, digital skills, both basic and advanced, will be essential to strengthen our collective resilience as a society: only digitally empowered and digitally capable citizens and digitally skilled workers can control their own destiny and be confident and assertive about their means, values and possibilities"[22].

The Spain Plan, through the set of reforms and investments that are part of it and, in particular, those aimed at digitalisation, automation and robotics, with the invaluable financial support of the Recovery and Resilience Fund, constitutes a unique opportunity in this area that cannot be missed.

---

20 Project on Tax Base Erosion and Profit Shifting (BEPS), and in particular its Action 1 Report on BEPS (BEPS), and in particular its Action 1 Report on BEPS (BEPS), and in particular its Action 1 Report on BEPS (BEPS). Tax challenges of the digital economy of 5 October 2015, https://www.oecd.org/ctp/beps-resumenes-informes-finales-2015.pdf. As well as the Interim Report on Tax Challenges arising from Digitisation of 16 March 2018. https://www.oecd.org/tax/beps/resumen-desafios-fiscales-derivados-de-la-digitalizacion-informe-provisional-2018.pdf.

21 https://eur-lex.europa.eu/legal-content/es/TXT/?uri=CELEX:52017DC0547.

22 European Commission, "Commission puts forward declaration on digital rights and principles for everyone in the EU", Press Release, Brussels, 26 January 2022, available at https://ec.europa.eu/commission/presscorner/detail/es/IP_22_452.

References

1. https://eur-lex.europa.eu/resource.html?uri=cellar:6785f365-1627-11ec-b4fe-01aa75ed71a1.0002.02/DOC_1&format=PDF
2. https://planderecuperacion.gob.es/
3. Propuesta de Decisión de Ejecución (PD): Consejo relativa a la Aprobación de la Evaluación del Plan de Recuperación y Resiliencia de España. https://ec.europa.eu/info/sites/default/files/com_322_1_es.pdf
4. https://eur-lex.europa.eu/legal-content/ES/TXT/?uri=CELEX:52020DC0103
5. https://www.nature.com/articles/s42256-021-00436-6
6. Proyecto sobre Erosión de Bases Imponibles y Traslado de Beneficios (BEPS, por sus siglas en inglés), y en especial su Informe relativo a la Acción 1 sobre los retos fiscales de la economía digital de 5 de octubre de 2015 (2015). https://www.oecd.org/ctp/beps-resumenes-informes-finales-2015.pdf
7. Informe intermedio sobre los retos fiscales derivados de la digitalización de 16 de marzo de (2018). https://www.oecd.org/tax/beps/resumen-desafios-fiscales-derivados-de-la-digitalizacion-informe-provisional-2018.pdf
8. https://eur-lex.europa.eu/legal-content/es/TXT/?uri=CELEX:52017DC0547
9. European Commission: Commission puts forward declaration on digital rights and principles for everyone in the EU. Press Release, Brussels, 26 January 2022. https://ec.europa.eu/commission/presscorner/detail/es/IP_22_452

Author Index

M. A. Grau Ruiz (Ed.): INBOTS 2021, BIOSYSROB 30, pp. 247–248, 2022.
https://doi.org/10.1007/978-3-031-04305-5